American Institute of Chemical Engineers

# Management Division Conference

Presentations at the
2008 AIChE Spring National Meeting

April 6-10, 2008
New Orleans, Louisiana, USA

**Printed from e-media with permission by:**

Curran Associates, Inc.
57 Morehouse Lane
Red Hook, NY 12571
www.proceedings.com

**ISBN: 978-1-60560-218-9**

**Some format issues inherent in the e-media version may also appear in this print version.**

ISBN: 978-1-60560-218-9



American Institute of Chemical Engineers
Proceedings
Three Park Avenue
New York, NY 10016-5991
Phone: 212-591-8100

www.aiche.org

American Institute of Chemical Engineers

Management Division Conference
2008

# TABLE OF CONTENTS

# 74a Synergy in Mergers of Petrochemical Companies within a Complex Considering Purchasing and Selling Advantage with Process Integration

Sung-Geun Yoon[1], Seung Bin Park[1], **Sunwon Park**[1], Jeongseok Lee[2], Peter M. Verderame[3], and Christodoulos A. Floudas[3]. (1) Department of Chemical and Biomolecular Engineering, Korea Advanced Institute of Science and Technology, Daejeon, South Korea, (2) Corporate R&D, LG Chem Ltd., Daejeon, South Korea, (3) Department of Chemical Engineering, Princeton University, Engineering Quadrangle, Princeton, NJ 08544

Mergers and Acquisitions, M&A, have been active in the petrochemical industry. However, the synergy created by the merger of petrochemical companies has rarely been studied, although it is the primary goal of a merger. This study deals with the merger of petrochemical companies located within one complex. Synergy considerations resulting from process network integration with fixed cost reduction and market power increase in upstream market (i.e., resource purchasing advantage) and downstream market (i.e., product selling advantage) are included. A novel mathematical model is formulated that represents the operation of a process network aiming at increasing the profitability of merged companies. The resource purchasing advantage and product selling advantage options are considered by means of various scenarios. The proposed model is applied to Korean Naphtha Cracking Center, NCC, companies in one complex. The results, presented in three case studies, demonstrate that a merger creates synergy primarily from the purchasing advantage and selling advantage options, while the process network integration which simply collects various processes can create little synergy.

Keywords: Merger and acquisition; Synergy; Market power; Process network; Naphtha cracking center; Petrochemical company; Petrochemical industry

**Permitting for fun & profit**

**Effective management of permitting & regulatory affairs in capital projects**

*Stephen Johnson*

*Flour Canada, Calgary, Alberta*

## ABSTRACT

For many project managers, permits and regulatory compliance are an area which presents far more liability than opportunity – a single missing permit can add huge cost and time overruns to a project, but even the most careful execution cannot guarantee success. Permit and regulatory requirements are often unclear, and can vary wildly from jurisdiction to jurisdiction, or over time within a given jurisdiction. As a further complication, regulatory specialists – when available – who really understand the relevant requirements are often relatively uninvolved in the schedule and budgetary concerns of any given project.

Nonetheless, these risks are fundamentally no different from other risks – they are subject to mitigation by effective analysis and planning. The focus of this session will be on cost effective ways to understand and manage permitting and regulatory compliance in large capital projects.

**Permitting for fun & profit**

**Effective management of permitting & regulatory affairs in capital projects**

*Stephen Johnson*

*Flour Canada, Calgary, Alberta*

## THE NATURE OF THE PROBLEM

A typical large capital project will require a variety of permits, from the large profile ones, such as the Environmental Impact Assessment (EIA), down to occupancy permits for site offices. The absence of any one permit can cause significant delays and additional costs to a project

Historically, with certain exceptions, permits have rarely been critical path drivers on reasonably well managed large capital projects. In recent years, however, compressed project schedules and longer approval cycles have moved some permitting activities onto the project critical path. This necessitates much more careful attention to the permitting process, and its many constituent activities.

## THE EIA & THE CRITICAL PATH

The Filing of the EIA is increasingly subject to a pair of independent forces which significantly complicate the filing process. Because of the compression of project schedules, and generally increased approval times, applications need to be filed much earlier in the design development process than in previous years, with much of the critical data still under development. At the same time, the level of review scrutiny by both governmental regulators and outside interveners (including both non-governmental organizations and lenders) has increased substantially. This means that much more care & focus must be taken on the clear establishment of the application design basis. In many cases, this needs to be a formal project deliverable, with all the associated review & alignment required to ensure that the design assumptions expressed in the EIA are

- Clearly spelled out
- Consistent with each other
- Technically credible

Clarity is essential on the documents being prepared as a basis for the EIA. Paradoxically, the EIA itself will disclose the absolute minimum data possible -however, it is essential for the parties involved in the development of the application to have the clearest possible understanding of the design at that frozen point in time. All unknowns should be spelled out and a determination made if they will be defined prior to submission of the application, and if not how the application will adres the lack.

The primary reason that clarity of the root documents is so essential is the stringent requirement for the highest possible level of consistency in the application. Errors of internal consistency within this document can easily lead to significant delays in approval and provide hostile interveners with significant justification to oppose the planned development. Lenders themselves may require review and/or approval of the application, and will be highly reluctant to back projects with poorly organized applications. This is particularly true of ***Equator Principles*** signatory lenders.

Technical credibility is also essential. If a project is, for instance to obtain unusually high environmental or safety standards, it is essential that the means by which those higher standards are to be met are described in sufficient detail to make a convincing case to an audience which will include regulators, members of the public, and experts in various technical disciplines. Falsehoods, exaggerations, misleading language and unsupported hand waving are to be avoided at all costs.

## QUASI-PERMITS

Many large capital projects significant increase the burden on local utilities (i.e. natural gas, water, electricity, etc.) and infrastructure (roads, rail, ports, etc.) and the service agreements with these entities work very much like permits, even though the entities involved may be conventional corporations. It is essential to obtain a solid understanding of the specific requirements each of these bodies may have of the project – an electric utility, for instance, may require long cycle (or, worse, open timeframe) approval of power distribution schemes as part of the supply agreement.

Municipalities – including master planned industrial communities – and other local jurisdictions will also have significant input to the project. Multiple local authorities – police, fire, sanitation, etc. may have completely independent approval requirements from those which the local elected officials may require. Further, the local residents – particularly indigenous peoples (where applicable) may have strong feelings about the project if they're not consulted effectively. It is essential to obtain a good understanding of the local stakeholders attitudes and concerns as early as possible.

## PITFALLS

The bulk of the permitting activity required for large capital projects already has reasonably robust mechanisms in place to ensure it happens fairly smoothly. However, there are some fairly significant gaps in the machinery, and the old saw still applies "People don't plan to fail, they fail to plan". The three most common are undocumented assumptions of responsibility, changes of jurisdiction, and developing design technology. Undocumented assumptions of responsibility are really simple failures of project execution – the project leadership didn't spell out who was responsible for a given permit or activity. The risks associated with changes in jurisdiction are fairly intuitive when the changes are gross , such as a company based in Europe building a facility in Africa – but even minor changes in jurisdiction can significantly alter the permitting activities needed – a facility whose original permits were clear sailing may face significant regulatory hurdles for an expansion if the region is now an environmentally sensitive or non-attainment area. The final – and in many ways the most sensitive are is how changing technology can change the permitting picture. Many of these may not be intuitive, such as the requirements that some countries have for jewelry dealing licenses for any import/export of precious metals – including industrial catalysts. Another example would be the classification of large process modules: a significant number of venues consider them to be buildings, and subject to local building codes, life safety requirements for access/egress, etc. – this is, of course, not typically true of stick-built process equipment or regular skid packages. This kind of change can impose significant constraints on the design of a unit, particularly if not discovered until late in the project. There's no magic wand to eliminate these risks – ultimately, there's no substitute for careful thought about the permitting picture at the front end.

## MITIGATING THE RISK

The first order of business on any large capital project is to assess your level of certainty regarding the regulatory environment, and the nature and diversity of the problems you will be facing. For instance, a chemical plant being built in a Saudi Arabian master planned industrial city like Jubail will – from a permitting standpoint – have a relatively straightforward path. The Royal Commission will, typically, be the place for all non-trivial permits, approval cycles and documentation requirements are well known, and the infrastructure interfaces will be reasonably simple. There are some geopolitical and logistic permitting risks, and schedule constraints with things like visas, travel during Ramadan, etc. – but the level of uncertainty is fairly low. On the other hand, a project in the developed world may face significant uncertainties regarding popular and NGO interveners – many members of the public are profoundly skeptical of the benefits of such heavy industrial facilities, and of the good intentions of the owners. If not handled well, they can put significant pressure on local officials to stop or delay projects, and a poorly prepared permitting plan increases the project exposure to such risks. So, once the level of understanding is assesses, the project can proceed with resolving the unclear elements – either by hiring a specialist or pursuing the knowledge within the task force. The use of consultants is a mixed blessing here – they often know the local regulatory scene very well, but can be fairly uninvolved in the overall project, and will often only work reactively – for instance, they will only do what is asked, without any attempt to understand what is actually needed.

In any case, once a reasonable level of understanding of the applicable permitting processes is obtained, the next thing that needs to happen is the establishment of a master permitting plan with a detailed permitting schedule and RACI/interface chart. This needs to identify all significant handoffs, milestones and – of course – the permitting critical path. It will also need to use significant milestone dates from the project (such as start of construction, etc.) to ensure it supports the project schedule. Like the project schedule itself, this needs to be a living document, monitored by the project leadership. Based on the plan, a 30 / 60 / 90 day look ahead plan can then be regularly extracted to watch for potential problems. Timely escalation of delayed permits can ensure that schedule and budget surprises are kept to a minimum. It is also essential that the permit plan be reviewed periodically, particularly prior to the main stage gates, such as financial sanctioning.

## A SUGGESTED STARTING POINT

There is no substitute for a well thought out, project specific plan, but a preliminary list of some items for consideration is attached below.

# PRELIMINARY PERMITTING CHECKLIST

## CORPORATE / PROGRAM / PROJECT LEVEL PERMITTING & APPROVALS

### "License to Operate" Requirements

- Business / Project entity licenses & permits – including in-country entity creation if required
- Feedstock import licenses, allocations and/or agreements
- Product disposition agreements / export permits & contracts
- Project capital goods export /technology/ licensor permits
- Product export licenses
- Employment / Residency permitting for expats and OCNs
- Customs brokerage agreements (Note: precious metals catalysts may require special impex handling)
- Insurance underwriting and/or classification approvals

### Environmental and related permits – permanent facilities

- Environmental Impact Assessment (Note: may have multiple stages and/or different versions for Regulators and financial underwriters)
- Fresh water usage / discharge permits
- Sea water usage / discharge permits
- Navigable water crossing permits (including for pipelines)
- Source / injection well permitting
- Air permits / emission credits
- Carbon permits / exchanges
- Cultural / historical / archaeological impact assessment
- Fisheries, wildlife, flora, wetlands, marine assessments
- Emergency Response plan
- Fire department / fire marshal approval
- Site security plan
- Noise plan
- Light pollution plan
- Area stakeholder meetings

### Land Use – Permanent facilities

- Land purchase, lease, or allocation
- Master-planned industrial area permits (common areas, sub-usufruct, etc.)
- Industrial co-operative agreements (Fire brigades, terminal operations, etc.)
- Right-of-way crossings
- Pipeline right-of-way acquisitions / permits
- Zoning approvals, variances & re-zoning requests
- Development plans

### Utility and Infrastructure access – Permanent facilities

- National / regional electrical grid access (may not be a permit, often works like one)
- Local power company/utility approvals and agreements
- Natural gas supply agreements
- Water and sewage agreements
- Landfill usage permits & agreements (including hazardous waste facilities)
- Solid waste storage and/or disposal facility agreements & permits
- Industrial utility supply agreements with individual neighbours
- Telephone / telecom / satellite / data services agreements & permits
- Radio service agreements, frequency allocations & permits
- Port or terminal service agreements, traffic plans & permits
- Rail service agreements & permits
- Road traffic permits and plan approvals

### Construction activities & temporary facilities

- Construction permits
- Construction camp occupancy / health / foodservice / sanitation permits
- Traffic and road usage permits
- Construction noise, dust, emissions, runoff permits
- Construction site fire protection / firefighting plans
- Occupancy / construction permits for temporary facilities

## UNIT LEVEL PERMITS & APPROVALS

### Buildings - permanent facilities

**NOTE**: In some locations equipment modules may be considered to be occupied buildings.

- Building inspection and/or occupancy permits
- Food services permits
- Medical facility permits
- Fire station permits

### Logistics, transport & impex

- Ship and other carrier vessel assignments & port berth assignments
- Road time allocations & oversize load permits
- Import / export licensing, customs brokerage, duty exemption filing & freight expediting
- Explosives permitting, importing, handling & storage
- Radioactive materials importing, handling & storage

### Unit level EPC Construction & temporary facility permits

- Disposal of construction waste, spoil, borrow, etc.
- Temporary electrical power permits
- Temporary water supply & disposal permits
- Special employment permits & licenses
- Jobsite sanitation / health / safety / food service & camp permits
- Jobsite kitchen / cafeteria permits
- Jobsite logistics / traffic plan

# PROJECT PORTFOLIO CAPITAL AT RISK AND THE ENTERPRISE PROJECT DELIVERY SYSTEM

For presentation to the AIChE Spring National Meeting,
April 6-10 in New Orleans, LA.
Authors: Pete Luan and John Wray
Westney Consulting Group
2200 West Loop South Sutie 500
Houston, TX 77027
713-861-0800

## Introduction

The impact of large, high risk capital projects in an energy firm's portfolio is frequently misunderstood and unappreciated. Often project delivery systems that were adequate for small or moderate size projects are totally inadequate for large projects that involve multi site international sourcing for engineering, material, fabrication and construction. Failure to recognize the mismatch and to upgrade the organization's project delivery system means that projects do not meet their business objectives such as: underperforming assets, late projects, and overrun budgets.

We present in this paper a methodology for evaluation of the project delivery system and its ability to support the firm's capital portfolio. Further we provide Capital at Risk models that notionally show a company's capital exposure for failure to upgrade the project delivery system.

## The Project Delivery System

A Project Delivery System is the methodology that an organization utilizes to take a business opportunity from concept to startup. This is not to be confused with a phase gate process which is but one component of a Project Delivery System.

A project delivery system is composed of three major elements:

- Process – Are the work processes and procedures in place within the organization to support the planning and execution of the project?
- People – Who will perform the work, their skill set, their knowledge, and their availability to work on the project? Do we have enough people with the appropriate skill set on the project?
- Organizational Governance – Is there appropriate oversight and support for the project team to ensure that the project is planned and executed in accordance with the requisite processes and controls?

The monetization of an opportunity utilizing these three components of the Project Delivery System is depicted in the graphic below.

**Figure 1 Project Delivery System Components**

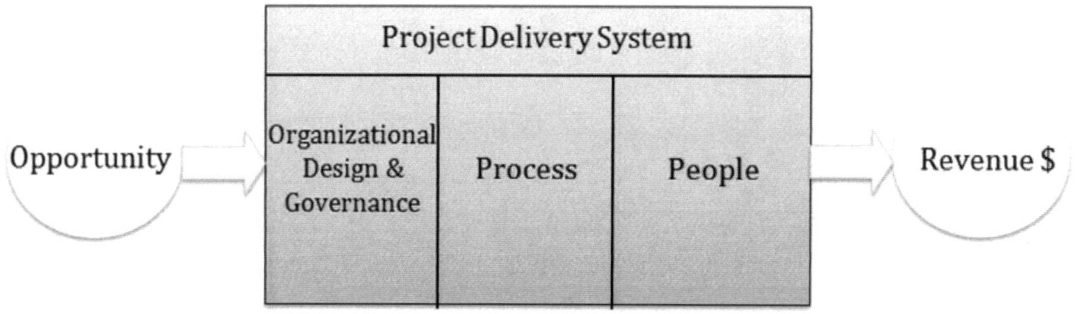


## Competencies

A competency is defined as an observable behavior or utilization of a requisite skill set. Westney Consulting Group's view is that there are twenty competencies that define the Project Delivery System. These twenty competencies are categorized into three major categories; strategic, tactical and performance management. The table below illustrates the three major categories and the associated competencies.

An organization must have these competencies to successfully plan and execute projects. Each of these competencies requires a business process, people with the requisite skills as well as executive organizational governance.

**Figure 2 Project Delivery System Competencies**

**Strategic**

- Partner Relationship Management
- Local/Government Issues Management
- Project Execution Planning
- Scope Management
- Value Improving Practice Implementation
- Risk Management
- Contracting Strategy
- Team Alignment & Effectiveness
- Interface Management
- Communication Management

**Tactical**

- Technology Management
- Design Management
- Procurement Management
- Construction Management
- Commissioning & Startup

**Performance**

- Safety/Health/Environmental Management
- Cost Management
- Schedule Management
- Quality Management
- Operability Management

Figure 3 shows the interaction between the three major Project Delivery System elements (Process, People, and Governance) and the three major competency categories (Strategic, Tactical, and Performance Management) where the twenty competencies reside.

**Figure 3 Project Delivery System Competency Model**

**PDS Competency Model**

| CATEGORIES OF COMPETENCIES | CATEGORIES OF PDS ELEMENTS: I Governance and Organization Design | II Work Processes, Methodologies & Systems | III People (Skills, Availability, Effectiveness...) |
|---|---|---|---|
| A Strategic Management Functions | Business Controls and Organization Strategy | Business Practices and Management Methods | Staffing Strategy |
| B Tactical Execution Functions | Alignment and Organizational Integration in Project Planning | Specific Execution Plans and Management | Providing Qualified and Experienced Personnel |
| C Project Performance Functions | Budgeting, Decision Making and Controls | Tools for Tracking, Forecasting and Controlling Projects | Performance Capabilities required for Project Success |

The organization's Project Delivery System is evaluated at the intersection points. As an example, for Contracting Strategy:

- Process – What are the processes in place to ensure a robust contracting strategy?
- People – Are there sufficient personnel with the appropriate skill set on the project team to develop a contracting strategy that supports the project objectives?
- Governance – Is there an existing executive oversight structure to ensure that people and processes are properly applied in planning and executing the project; and results are aligned with business objectives?

This same process is followed for the remaining nineteen competencies. From the analysis a series of findings are developed. The graphic below illustrates the typical results for a firm that is transitioning from executing small domestic projects, \$10MM - \$100MM, to a portfolio where the majority of the projects are international in scope and in the \$500MM - 1000MM++ range.

Note that the evaluation of an organization's Project Delivery System is always done in the context of their project portfolio. A firm that can execute small U.S. based projects will not have the competencies to successfully deliver internationally focused megaprojects. Figure 4 illustrates the results for the risk management competency. Clearly the only way the organization will deliver projects consistently is by having strong processes and personnel with the right skill with a strong governance structure.

**Figure 4 Sample Risk Management Competency Evaluation**

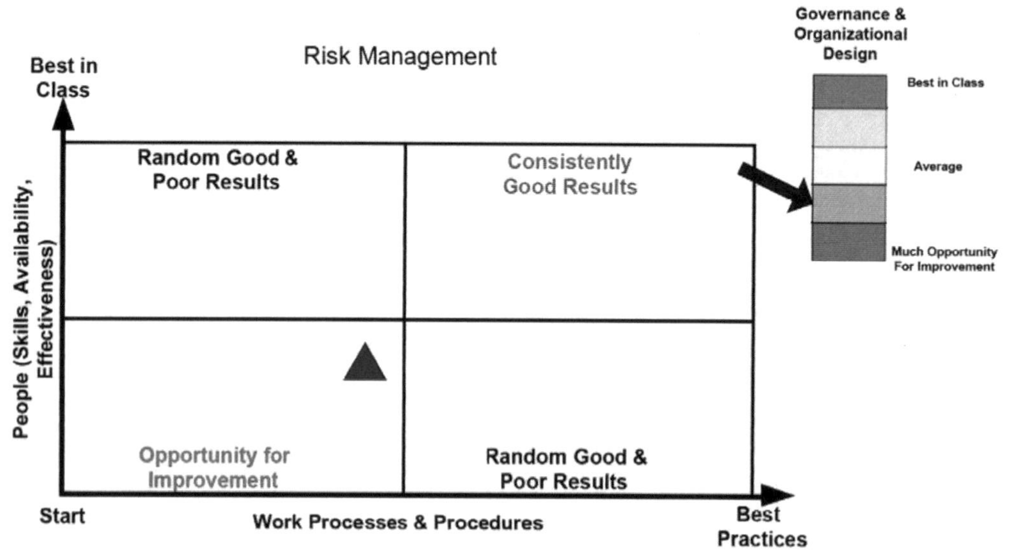


This process is followed for all of the rest of the competencies and a picture emerges as to the ability of the firm to support the project portfolio.

**Figure 5 Notional Firm - Project Delivery System Competency Evaluation**

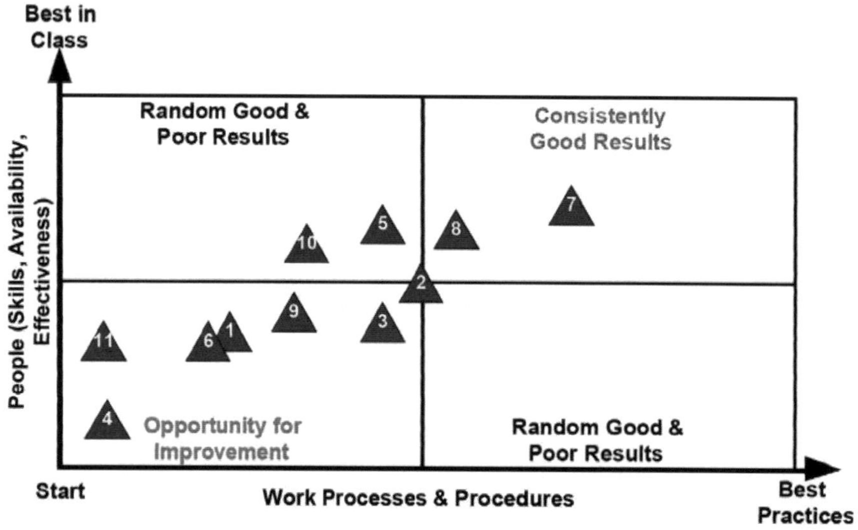

Based on the findings a 1 to 5 rating is assigned based on the capability of the firm's Project Delivery System to support the project portfolio: One being the lowest and five the highest.

## Portfolio Analysis

As previously discussed, the Project Delivery System must be sufficiently robust to support the portfolio and minimize risks. The next step is to categorize each opportunity in the portfolio. Through a proprietary methodology utilizing various criteria a rating of I to V is assigned to each project. A Type I project is the least challenging and a Type V the most challenging. Figure 6 is an example of portfolio categories developed for a medium size energy company with both domestic and international projects. The descriptions must be adjusted to match each firm's portfolio.

**Figure 6 Project Portfolio Categorization**

| Project Type | Risk | Description |
|---|---|---|
| **Type I** | **Low Risk** | CAPEX less than $5MM, very simple scope, very mature technology, and client is a U.S. based long time partner. |
| **Type II** | **Some Risk** | CAPEX $5MM to $10MM, scope is well understood, international installation, well known technology, and there is some experience with the non U.S. based client. |
| **Type III** | **Moderate Risk** | CAPEX is $10MM to $50MM, scope is an integrated unit, some work is to be done internationally, have some experience with the technology. |
| **Type IV** | **High Risk** | CAPEX is $50MM to $200MM, scope includes multiple integrated units, engineering and construction will be done internationally, and firm has limited knowledge the technology. |
| **Type V** | **Highest Risk** | CAPEX is greater than $200MM, scope includes multiple integrated units, engineering and construction will be done internationally, and firm has never used the technology. Owner has a challenging construction schedule. |

## Capital at Risk (CaR)

Capital at Risk, or Cost at Risk (CaR), is defined as the difference between the worst case outcome for project cost and the project authorization amount. The worst case amount is usually defined as a P80 value or a value that has an 80% probability of not being exceeded. Based on Westney Consulting Group's experience, a project authorization value is typically less than P50, which means that the project has higher than a 50% probability of exceeding the authorized amount. The figure below illustrates the concept of CaR:

**Figure 7 Project Capital at Risk**

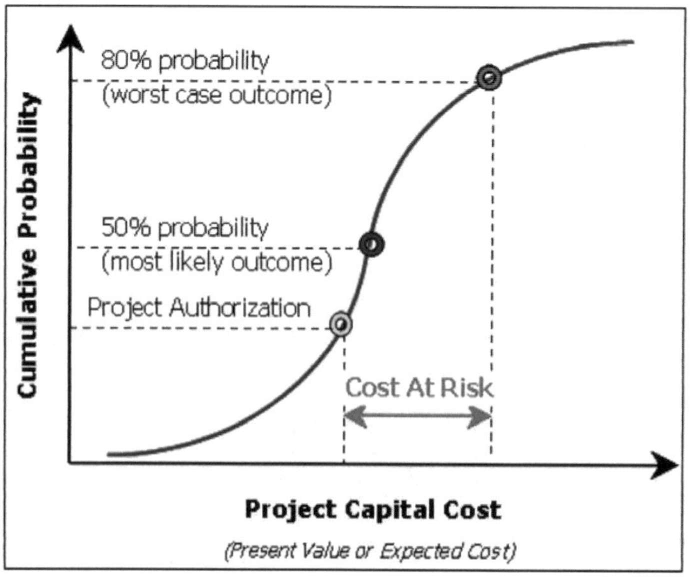


For a given Project Delivery System rating, the more complex the project (a Type V versus a Type I) the higher the Capital at Risk will be for the firm. Alternately, for a given organization's portfolio, the lower its Project Delivery System rating, the higher its Capital at Risk.

Higher Capital at Risk translates into cost overruns, schedule delays, poor asset performance, etc. This concept is demonstrated in the Figure 8. The vertical axis is a plot of the percent Capital at Risk of the base capital cost estimate. The horizontal axis is the Project Delivery System rating of the organization.

This notional firm depicted in Figure 8 has a PDS ranking of 1. The majority of its portfolio is relatively simple Type 1 projects. However, the firm has several more complex Type III and Type IV projects. Failure to upgrade its Project Delivery System will mean that a substantial portion of its capital will be at risk as the increasing percentage of the CaR is applied against an ever increasing capital project cost!

However, if the firm can transition to a PDS ranking of 3 it will be able to execute more complex, higher capital, and presumably higher ROI projects without an unmanageable Capital at Risk.

**Figure 8 Capital at Risk versus the Project Delivery System**

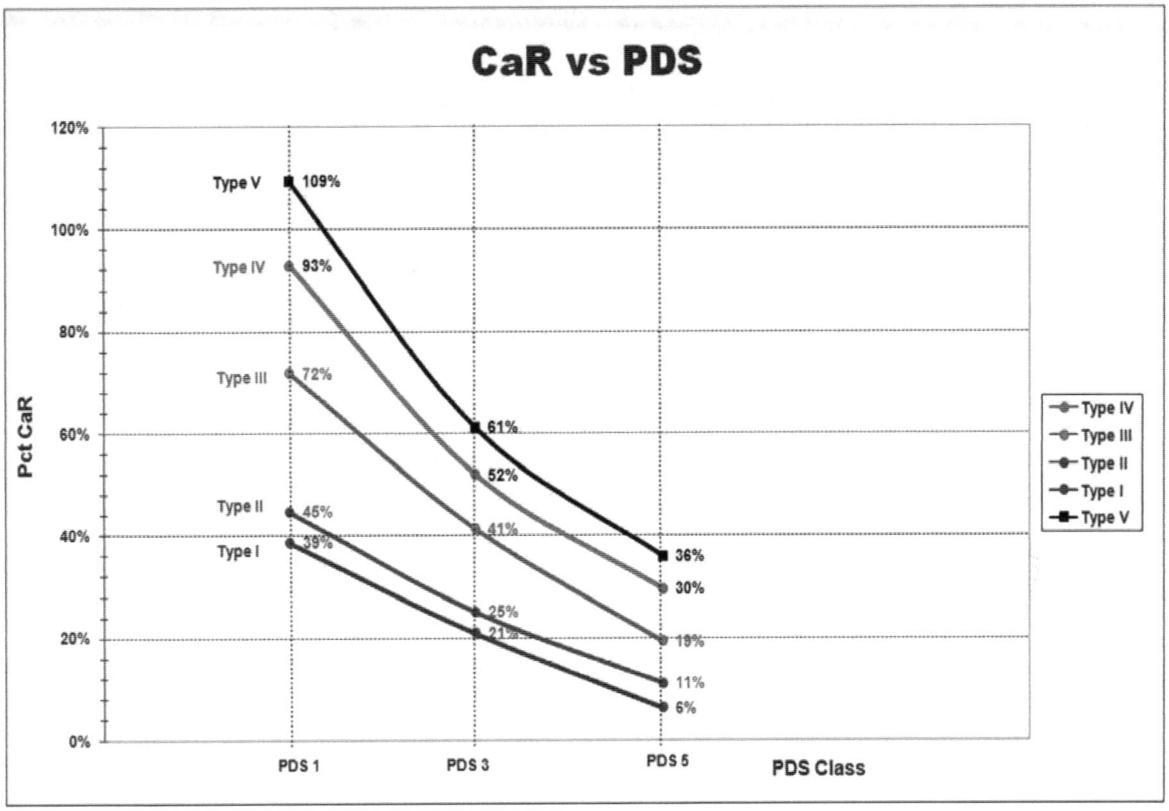


## Conclusion

As organizations strive to execute larger projects in a challenging environment they must upgrade their Project Delivery System to meet the increasing project complexity. Failure to do so means the firm puts substantial Capital at Risk as is demonstrated by project schedule delays, cost over runs, poor asset performance, etc.

What we have discussed in this paper is a methodology to a

- Assess the current state of a firm's Project Delivery System based on a set of twenty competencies
- Analyze and categorize the complexity of the project portfolio
- Determine notionally the organization's Capital at Risk based on the existing Project Delivery System in its project portfolio.

Authors

*At Westney Consulting Group Pete Luan works with business unit leaders, CEO's and COO's to ensure that capital projects meet their business goals. A former officer in the US Army, he gained extensive experience managing large capital projects for BP Amoco in the Middle East and Latin America. He has held management positions with several management consulting firms and is a certified Project Management Professional. He holds BS and MS degrees in Mechanical Engineering (with high honors) from Rice University and is a graduate of Harvard Business School Executive Education program.*

*John Wray is Vice – President and a Senior Executive Consultant with Westney Consulting Group. He specializes in project management best practices including lessons learned and team alignment. John held various senior management positions with DuPont, focusing on engineering, plant operations and maintenance, and project management. He holds a BS in Electrical Engineering from the University of Texas and is certified as a Project Management Professional as well as a Myers Briggs consultant.*

# Managing Uncertainties in Project Management of Hydrocarbon Projects

**Syamal K. Poddar Ph.D., P.E., Fellow AIChE**
**Poddar & Associates**
**P.O. Box 890461**
**Houston, TX 77592-0461**

## Abstract:

The effective technique and procedure for successfully managing any project, whether it is a complex mega project or a relatively simple & small project relating to any hydrocarbon industry, or for that matter for any industry sector, always involve several critical steps.

The interdependence of project management technique and the various types of industry standard contract types are discussed with reference to respective potential risk tolerance characteristics of each type of contract

Finally the applicability of earned value analysis in project management as an important measure of project status is elucidated.

## Critical Steps of given hydrocarbon project:

1. **Study Phase:** This is the very initial phase of the development of any project. There are usually three progressive and sequential phases of the study phase. They are pre-feasibility, feasibility and detailed feasibility phase. This phase drives any given project from conceptual stage to a well defined project stage, including a clear picture of the 3W's and 1 H (**W**hy, **W**hat, **W**hen & **H**ow)

2. **Engineering Phase:** There are two sequential parts to this phase.

They are Front End Engineering (FEE) & Detailed Engineering (DE). The FEE phase often overlaps with the detailed feasibility study. In this phase the project definition, including Scope of Work (SOW) as well as the technology selection together with the Basic Engineering Design (BED) or often called Front End Engineering Design (FEED) package, are completed. Supported by other engineering disciplines in this phase the process engineers play a leading role. In this phase the BFD and PFD are developed

In the detailed engineering phase the FEED package is broken up into various engineering discipline oriented packages. The detailed engineering is done where the detailed plant layout, engineering for the entire plant, including foundation, structure, equipment design and sizing, piping, instrumentation and control, electrical, are completed. All P&ID's are completed and detailed HAZOPs are done with direct involvement of the representatives of all stake holders of the project

3. **Procurement Phase:** In this phase procurement of all equipment is done in a systematic way, delivered to the project site per schedule . This particular phase involves a significant level of logistics challenges. This phase is not necessarily in sequence with the Detailed Engineering phase. The procurement initiatives for some of the "long lead", high demand items are done in parallel with the detailed engineering phase.

4. **Construction Phase:** This phase is a crucial phase for any EPC project and requires a significant level of management expertise to make it a success. Besides transforming the detailed engineering document to a reality it also includes start up commissioning and hand over of the project to operating groups.

Success of any given project is a combination and a summation of the success of each of these phases. Such individual phase-wise successes or the success of an entire project depends on how carefully the risks are managed and mitigated. The risks arise due to potential uncertainties in two key elements of a project execution. These are: Cost and Schedule which most often are very much interlinked. Besides that there are other risks e.g., risk associated with contract type and often environmental risks, particularly, for projects involving construction phase.

The ability to influence cost and/or schedule for any given project is very much dependent on at what phase of activities the project execution is situated with reference to the life cycle of the project. The more mature the project execution stage the less the possibility of influencing the overrun in cost and/or schedule of the entire project.. Such dependencies are qualitatively elucidated in figure 1 below.

## Contract Types and associated Risks:

Basically there are four types of contracts. They are:

- Lump–sum or Fixed price (Hard – money)
- Reimbursable
- Reimbursable with fixed fee
- Unit price

Contract types and associated risks are summarized below in Table 1 and qualitatively shown in Figure 2:

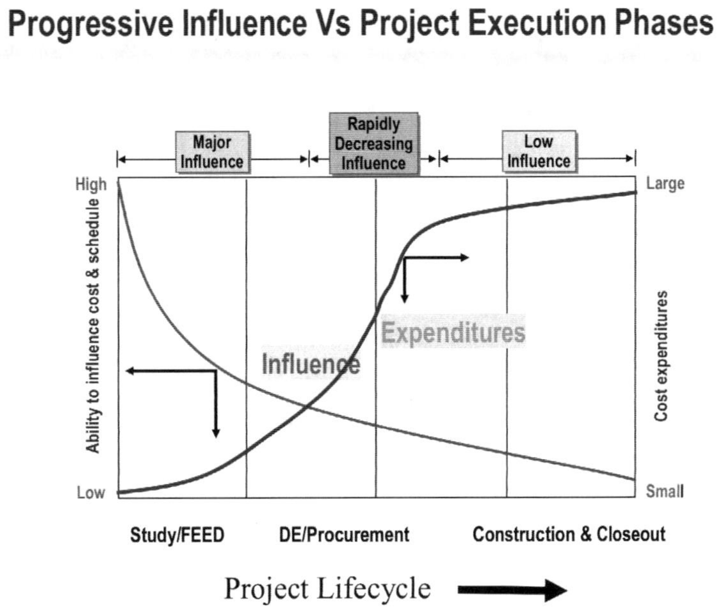


**Figure 1**

| Criteria | Lump sum | Reimbursable | Reimbursable with fixed fee | Unit price |
|---|---|---|---|---|
| Pricing | Highly competitive | Competitive | Competitive | Competitive |
| Contractor risk of Cost Overrun | High | Lower | Moderate | Low |
| SoW definition | Detailed definition, fixed scope | General definition, variable scope | General definition, variable scope | Semi-detailed definition variable scope |
| Client risk of cost overrun | Lower | Higher | Moderate | Moderate |

**Table 1**

There are a number of ways any given contract, particularly Lump Sum type, could be incentivised to bring a better alliance between the contractor and owner, and thereby bringing optimum performance with potential overall positive impact on the project cost and/or schedule. Such incentive plans are qualitatively exemplified in figure 3 below

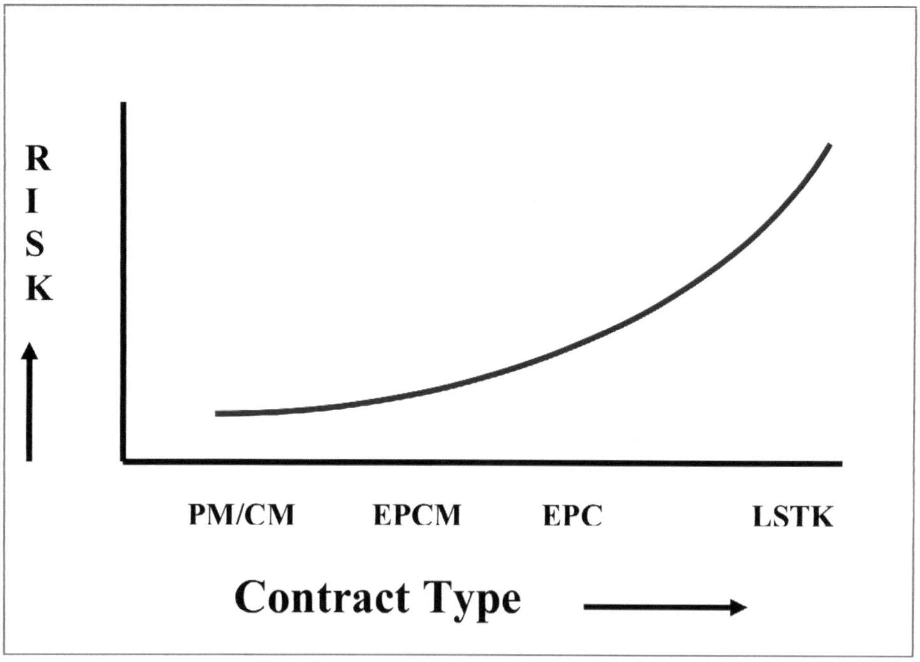


Figure 2

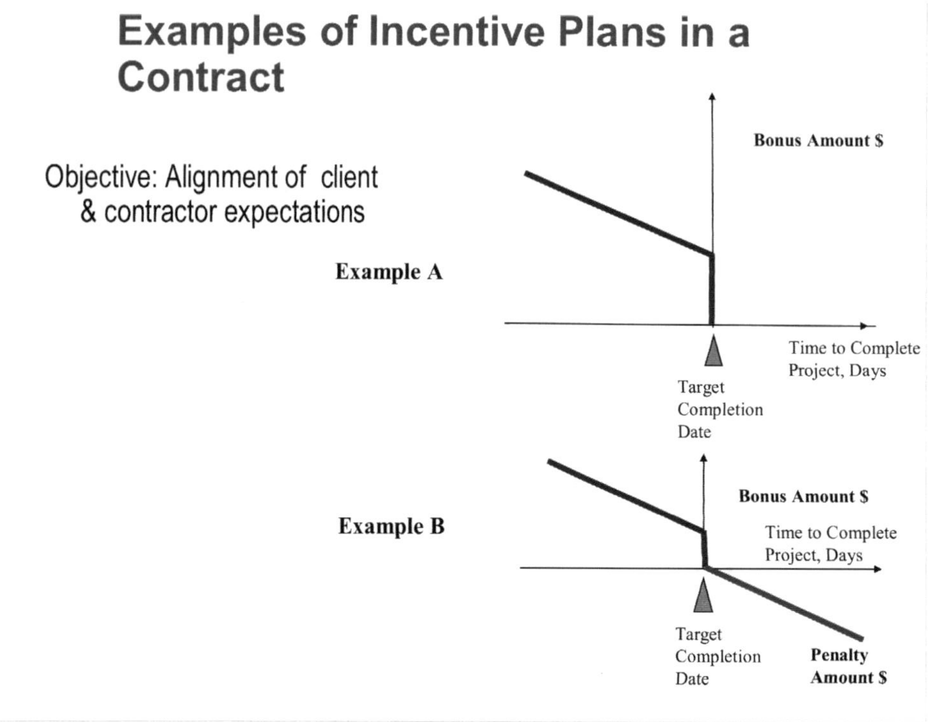


Figure 3

## Means to Minimize the Risk of over run of Costs and/or Schedules:

Defining the Project Scope clearly and with details and
Effective Project Planning ( e.g., IPA's FEL approach)
Well established QA & QC
Effective control of Scope of Work (SoW)
Effective 360 degree communications
Reliable suppliers of
- materials & equipment
- fabrication and
- subcontracts

Effective Project Management Team (PMT) guided by a Project Manager with superb leadership Skills

**Leadership Skills:**
- Planning
- Cost control
- Interface management
- Interpersonal skills
- Sensitive to individual s and cultural diversities
- Business acumen (contracting & contract language)
- Understand and anticipate (proactive and not reactive)
- Represent his company as well as the client
- Understanding of project expectations and exceeding them
- Cultivating Team Work culture
- Building **TRUST** and delegating responsibilities
- Cultivating a strong sense of accountability

Incorporation of reliable and dependable project management tools and procedures

## Earned Value Analysis:

This is a powerful analytical tool to determine the true progress status of a given project during execution stage. When projects are on schedule the spend plan approach is relevant to monitor to project status. For projects which are behind schedule such monitoring provides a false picture of the project status. In order to monitor the actual progress of any complex large project, rigorous tracking of cost and schedule are essential.

At a given time during project execution such an analysis integrates three distinct pieces of information in dollars spent or in hours spent. These are: 1) work planned to be accomplished (Planned Value), 2) actual spending or actual cost and 3) the value of the work performed with reference to the baseline budget which is defined as Earned Value.

Analysis of these numbers provides a real picture of any project at a given time ( a "snap shot" of a project execution status) with reference to under run, over run or on track.

## Concluding remarks:

To successfully execute a project, small or large, the understanding of key execution steps of the life cycle of a project is extremely important. Such understanding results in a clear picture of risks of various steps which are very much influenced by the uncertainties associated in each of these steps. A probabilistic analysis of each of these risks using commercially available Monte Carlo based software could quantify each of these uncertainties with a given confidence level. A careful approach to manage these risks is an essential part of successful project management. A successful project execution delivers two important end-results. They are: **profit** and most importantly **client satisfaction**...

.

## 74e Analyzing Mobile Service Provider's Strategy in Providing Mobile Digital Content and Services in Next Generation Networks

**Peng-Ting Chen**, Department of Business Administration, I-Shou University, No.1, Sec. 1, Syuecheng Rd., Dashu Township, Kaohsiung County 840 Taiwan R.O.C., Kaohsiung, Taiwan and Joe Cheng, Telcordia, Hsinchu, 840, Taiwan.

By having the characteristics of convenience, personalization, flexibility, security, position, ubiquity and dissemination, handheld mobile devices connected to telecommunications networks are predicted to become a critical way to gain access to the to the increasing amounts of digital content and services. The 3G technology and next generation network (NGN) have broken through the limitation of bandwidth allowing a higher transmission rate and more complex digital content and services interactions. However, the revenue of current mobile data services in Taiwan is only 7.7% of ARPU, compared to 26.6% in Japan, indicating that Taiwan mobile digital content and services still not yet universal. The lack of attractive and innovative digital content and services is the mobile service providers' present predicament. This study uses Delphi method to construct the hierarchy model while mobile service provider providing digital content and services in NGN. Then, the analytical hierarchy process (AHP) is conducted to get the relative weighting of each factor. Base on the result of AHP, this research introduces fuzzy multiple criteria decision making (FMCDM) method to determine the most practical business strategy for mobile service provider while providing mobile digital content and services. Due to the penetration rates for mobile phones reach saturation and the competition of the conventional voice service fees, many mobile service providers have faced the gradual decrease in average revenue per user (ARPU) of the contracted accounts and the shrinkage of their profits. The key success factors and business strategy suggestions we proposed are able to assistant mobile service providers while constructing their strategy in providing digital content and services.

# Hurricane Timeline Tools: Objective Guidance for Triggering Your Hurricane Response Plan

Chris Hebert, Bob Weinzapfel, ImpactWeather, Inc., Houston, TX

Hurricanes represent the greatest threat to business continuity for any organization located within several hundred miles of the coast. A hurricane can completely destroy hundreds of square miles of homes and businesses in a matter of hours. And it's not a question of whether a specific location will be impacted, it's a question of when. Because of this significant threat, every organization located in a hurricane-prone area has some type of hurricane response plan. Some are well thought-out, identifying each step or phase along the path to a complete shutdown of operations.

A popular type of hurricane action plan involves a phased approach to a developing threat. Each phase involves certain "trigger points" or thresholds that must be reached at which time action must be taken to ensure the safe shut-down of operations and/or the repositioning of personnel and equipment to a backup location. One of the main problems with using the phased approach is the setting of objective guidelines to trigger phase escalation. Some hurricane action plans use what appear to be objective guidelines, such as advancing the plan when a hurricane is within "X" miles of a location, or if a hurricane enters a particular body of water like the Caribbean Sea or Gulf of Mexico. The problem with this approach is that no two hurricanes are created equally. Decisions made using such guidelines are prone to errors, resulting in either acting too early or not early enough, given the developing threat. In order to provide better objective guidelines, a set of Hurricane Timeline Tools has been developed. These include:

- Hurricane Risk Indicator
- Probability of Wind Impact
- Worst-Case Scenario
- Hurricane Severity Index

## Hurricane Risk Indicator

Modern satellite and improved long-range computer model forecasts allow the meteorologist to better identify areas of potential tropical cyclone genesis and movement. Often, potential development can be predicted anywhere from several days to over a week before a tropical cyclone forms. A careful analysis of current and predicted surface and upper-level winds in advance of a tropical disturbance can allow the meteorologist to estimate the chances of the disturbance developing into a hurricane over the coming week. Once a potential hurricane development threat is identified, its most likely track is calculated. Locations in or near the path of this potential hurricane are identified. It is then communicated to these locations that there is a chance of a possible hurricane threat over the coming week. This is the Hurricane Risk Indicator – an early heads-up of a potential hurricane risk over the next seven days.

All tropical disturbances that originate in the Atlantic Basin are identified and tracked throughout the hurricane season. If a disturbance is judged to have a fair chance of developing into a tropical storm or a hurricane, and if the disturbance is projected to track in a direction that could threaten any land areas, then the disturbance is classified as a "Significant Disturbance" and a 7-day forecast track is prepared. The meteorologist identifies the points along the forecast track where tropical storm and later hurricane intensity could be reached at the earliest. Knowing the average 5-year forecast track error, a cone of uncertainty can be generated for the 7-day forecast track. The cone begins at the point where the meteorologist estimates that hurricane strength could first be reached and it extends out to day seven. For example, if the meteorologist determines that there is little or no chance that a system could become a hurricane prior to the 72-hour forecast point, then the cone of uncertainty will begin at the 72-hour forecast point and extend through day seven. Regions within this cone of uncertainty are said to have a "Positive" Hurricane Risk Indicator, meaning that there is an elevated risk of a hurricane impact within the next seven days.

The Hurricane Risk Indicator is a tool that allows a business to advance its hurricane action plan beyond the initial state of readiness that exists at the start of the hurricane season. Often a potential hurricane risk can be identified days before a tropical disturbance is classified as a tropical depression, tropical storm or hurricane by the National Hurricane Center. The early warning can help a business to deal with a potential hurricane risk long before it becomes an imminent threat, allowing for a smooth and timely implementation of its hurricane action plan.

## Probability of Wind Impact

With a potential hurricane risk identified, the next step in most hurricane plans involves estimating the chances of a hurricane impact and the possible timing of such an impact. Once a tropical cyclone (tropical depression, tropical storm, or hurricane) develops, the probability of any one of its wind fields impacting a given location can be calculated. This is called the Probability of Wind Impact (PWI). The procedure for calculating the PWI is described below.

The PWI assigns a probability or percent chance that a location will experience winds of 25, 39, 58, 74 or 100 mph with the passage of a tropical cyclone. It utilizes the latest forecast for the track and the projected size of the various wind fields. A 75% probability cone is derived from a 5-year mean track error. Knowing the distance across the 75% probability cone and the size of the wind field as the tropical cyclone passes a location, the probability of a location receiving a particular wind level can be calculated.

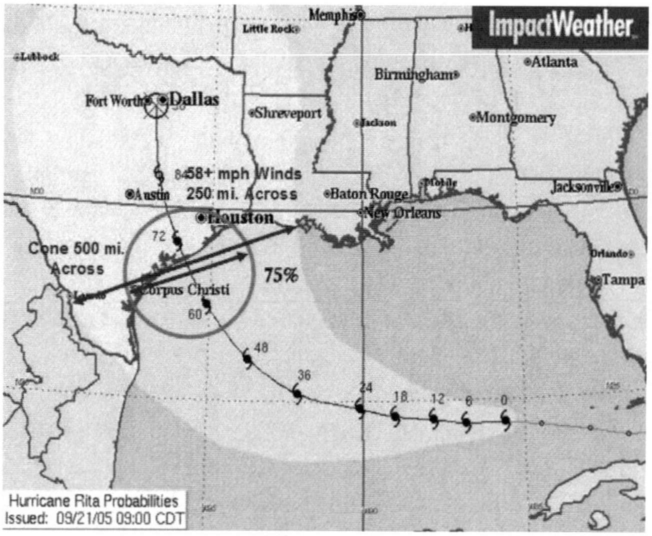


**Figure 1**

**Figure 1** is an example showing Hurricane Rita approximately 72 hours from landfall. Note that at 72 hours, the hurricane force winds are 160 miles across and the cone is 500 miles across. Using these data, a Gaussian (normal) distribution curve can be assumed (**Figure 2**). The white upward-pointing arrow is assumed to be the forecast track. Since the forecast cone in this example contains 75% of the possible latitude/longitude positions, 12.5% of the probability is assumed to lie to the right of the probability cone and 12.5% is assumed to lie to the left of the probability cone. Knowing a specific location with respect to the forecast track, the percentage chance that the location will be impacted by any particular wind field can be calculated.

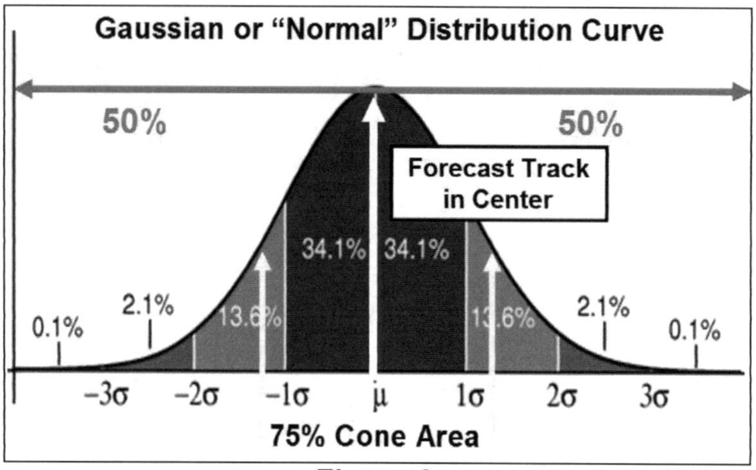


**Figure 2**

**Factors That Can Change the PWI at a Specific Location:**

1. A change in the projected size of the wind radii
2. A change in forecast confidence
3. Length of time until landfall
4. A shift in the forecast track toward or away from a point

Knowing the probability that tropical storm force or hurricane force winds will impact a location is key information that risk managers can use in the decision to escalate (or de-escalate) a phased hurricane action plan. However, risk managers typically run into difficulty when trying to implement such probabilities into their hurricane action plans. Which wind field should be used (39, 58, or 74 mph)? What values typically exist 3, 4 or 5 days from landfall? A study of PWI values for various hurricanes five days from landfall indicates that the best wind field to use is the 58 mph (50 kt) wind field. Using the tropical storm force wind field yielded values that were very high for large areas of the coast, even from three to five days before landfall. These high values didn't change very much from day to day. The reverse was true for the hurricane wind field probabilities. Many hurricanes have very small radii of hurricane force winds, leading to very low probabilities from three to five days before landfall. The 58 mph or greater wind field represents the part of the hurricane which will produce significant damage upon landfall. It is typically large enough such that probabilities change significantly each day from five days before landfall.

## Worst-Case Scenario

Tropical cyclone forecasting is not an exact science. The average 5-day track forecast error over the past 5 years is near 400 statute miles (**Figure 3**). Basically, the 5-day error can be identified by a circle 400 statute miles in radius drawn from the projected 5-day track point. There are two types of track error – cross-track and along-track. A track forecast can be off to one side or the other (cross-track). But a track forecast can also be wrong because the tropical cyclone moves either faster or slower than forecast (along-track error). Considering that the average 5-day error is 400 statute miles, and the average forward speed of a tropical cyclone is from 12-14 mph, it's possible that the tropical cyclone could impact a location over 24 hours earlier than indicated in the 5-day forecast. This earliest possible arrival time is the Worst-Case Scenario.

The Worst-Case Scenario (WCS) determines the earliest likely arrival times of a tropical cyclone's wind field. There are several assumptions when calculating the WCS for a specific location:

- The tropical cyclone takes a direct path toward the location
- The tropical cyclone moves at the maximum projected speed of advance
- The tropical cyclone is the maximum projected size and intensity at landfall

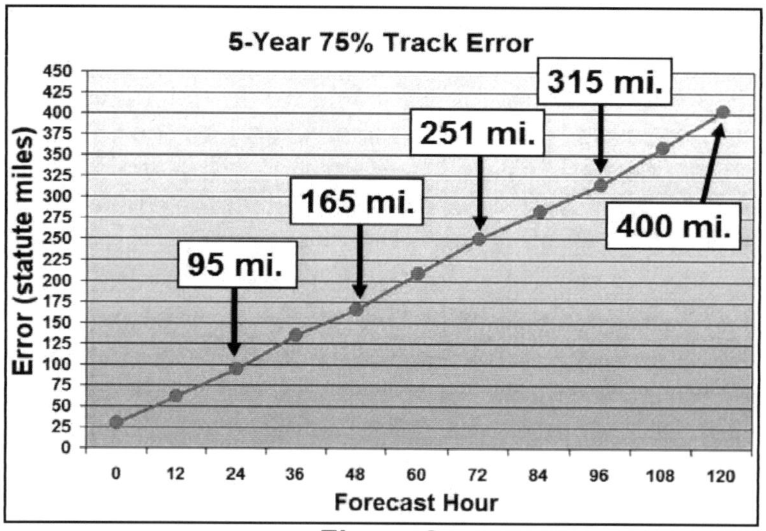


**Figure 3**

## Hurricane Severity Index

Hurricane Katrina devastated the coasts of Mississippi and Louisiana as a Saffir-Simpson Category 3 hurricane in 2005. Katrina produced a larger and more extensive storm surge and considerably more damage than did Category 5 Hurricane Camille in 1969. Coastal residents found it difficult, if not impossible, to believe that such damage could be produced by "only a Category 3 hurricane." Less than four weeks after Katrina made landfall, the large and powerful Hurricane Rita produced one of the largest and earliest coastal evacuations in history. Three days before Rita made landfall near the Texas/Louisiana border, Houston, TX, the fourth largest city in the United States, was nearly a ghost town. It was difficult to find even basic supplies between 72 and 96 hours before Rita made landfall. Clearly, the Saffir-Simpson Hurricane Scale is not adequate to measure a tropical cyclone's true destructive potential. More importantly, the Saffir-Simpson Hurricane Scale is inadequate in helping to predict possible community reaction to a developing threat. There is a clear need for a new hurricane intensity/strength scale which takes into account not only a tropical cyclone's maximum sustained winds but also its wind field size. The Hurricane Severity Index (HSI) was developed to address this need.

The HSI is a 50-point scale. Of the 50 points, 25 points are contributed by a tropical cyclone's intensity and 25 points are contributed by the size of its wind field. A tropical cyclone's intensity points are based upon its maximum sustained wind between 30 knots (35-mph) and 150 knots (175 mph). A 30-knot (35-mph) tropical depression receives 1 intensity point, while a 150-knot (175 mph) hurricane receives the full 25 intensity points. The scale is exponential (blue curve in **Figure 4**) and is based upon the known relationship of wind speed to the force exerted on an object. Wind force is related to the square of the wind velocity. This relationship is depicted by the red curve in **Figure 4**. Tropical cyclone intensity points are more heavily weighted toward hurricane-force winds or greater.

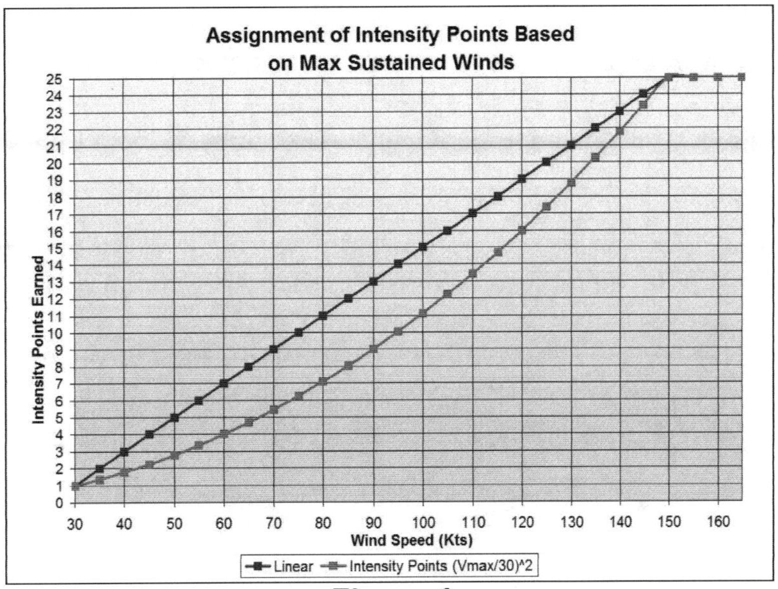


**Figure 4**

Determining how to assign points based on wind field size was a more difficult problem than assigning intensity points. The first step was to conduct a thorough study of past tropical cyclones. The National Hurricane Center's extended best-track dataset of all named storms from 1988 through 2005 was utilized, removing all extratropical cyclones, tropical lows and tropical waves. From the dataset, the 35, 50, and 65-knot wind radii for nearly 3800 data points were recorded. In addition to the three standard wind radii, an 87-knot (100-mph) wind radius was also included. ImpactWeather meteorologists have been forecasting the 100-mph wind radius for the past 10 years to better define the core of a strong hurricane in their day-to-day site-specific forecasts of wind impact for our clients. The 100-mph radius provides a good estimate of the size of a hurricane's core of intense winds, an area where extreme structural damage is prevalent.

Since 87-knot wind radii were not part of the NHC extended best-track database, those radii were measured from the detailed H*Wind Hurricane Research Division post-storm analyses for the 2001-2005 seasons. From these measurements, a formula was generated to estimate the 87-knot wind radii for any hurricane based on the known 65-knot wind radii and the maximum sustained winds using the following multiple linear regression equation.

If $V_{\max} \geq 87$ knot and $\mathrm{Re}_{87}$ is unknown,

$$\mathrm{Re}_{87\,estimated} = 0.5683\,\mathrm{Re}_{65} + 0.0792V_{\max} - 9.5383$$

Note that this equation estimates the effective radius of the 87-knot wind field based on the known effective radius of the 65-knot wind field and the maximum sustained winds of the hurricane. Once all wind radii were either measured or estimated, the average coverage of the 35, 50, 65 and 87-knot winds in every storm from 1988 through

2005 was calculated. This provided a baseline for determining whether a tropical cyclone's wind field was below average, average or above average.

A ranking system was devised to assign wind field size points based upon two primary factors:

- Areal coverage of wind radii compared to a historical range of tropical cyclones
- The known relationship between wind speed and the force exerted on an object

Doubling the wind speed results in a quadrupling of the wind force on an object. Therefore, size points were weighted heavily toward the stronger wind radii based upon this relationship. **Table 1** shows how the 25 size points are awarded:

| Wind Radii | Size Point Range |
|---|---|
| 35 kts | 1-3 |
| 50 kts | 1-4 |
| 65 kts | 1-8 |
| 87 kts | 1-10 |

**Table 1**

A tropical cyclone's wind field can be highly asymmetrical, making it difficult to compare wind field coverage between tropical cyclones. To standardize the wind radii of all tropical cyclones in the dataset, an effective radius (Re) for each wind threshold was computed. The effective radius is defined as follows:

$$\mathrm{Re} = \frac{1}{2}\sqrt{(RNE^2 + RSE^2 + RSW^2 + RNW^2)}$$

The effective radius defines the radius of a circle that has the same areal coverage as the tropical cyclone's wind field. For example, a hurricane may have a 65-knot (75-mph) wind field that extends 30 nautical miles (nm) only in the northeast and southeast quadrants. Winds in the southwest and northwest quadrants are below 65 knots. Using the equation above would yield an effective radius of 21.2 nm. That is, a circle with a radius of 21.2 nm would have the same areal coverage of 65-knot winds as a hurricane with 65-knot winds extending out 30 nm from the center in the northeast and southeast quadrants. Effective radius provides an easier way to compare one tropical cyclone's size characteristics to another.

To calculate how many size points a tropical cyclone would receive for each of the four wind radii described earlier, all of the effective wind radii values for each of the four wind fields were computed. The top 5 percent of the wind radii were then removed, as there were clearly a few outliers with very large wind fields that would have skewed the results toward high ranges. Similarly, the lower 5 percent of the wind radii were removed to eliminate any unusually small tropical cyclones. With the upper and lower 5 percent of wind radii removed, a realistic range for each of the four wind radii was derived. These ranges were divided equally according to the number of size points possible in **Table 1**. The 35-knot (39-mph) wind radius range was divided into three equal ranges, the 50-knot (58-mph) into four equal ranges, the 65-knot (75-mph) into eight equal ranges, and the 87-knot (100-mph) radii were divided into ten equal ranges. These values may be found in **Figure 5**.

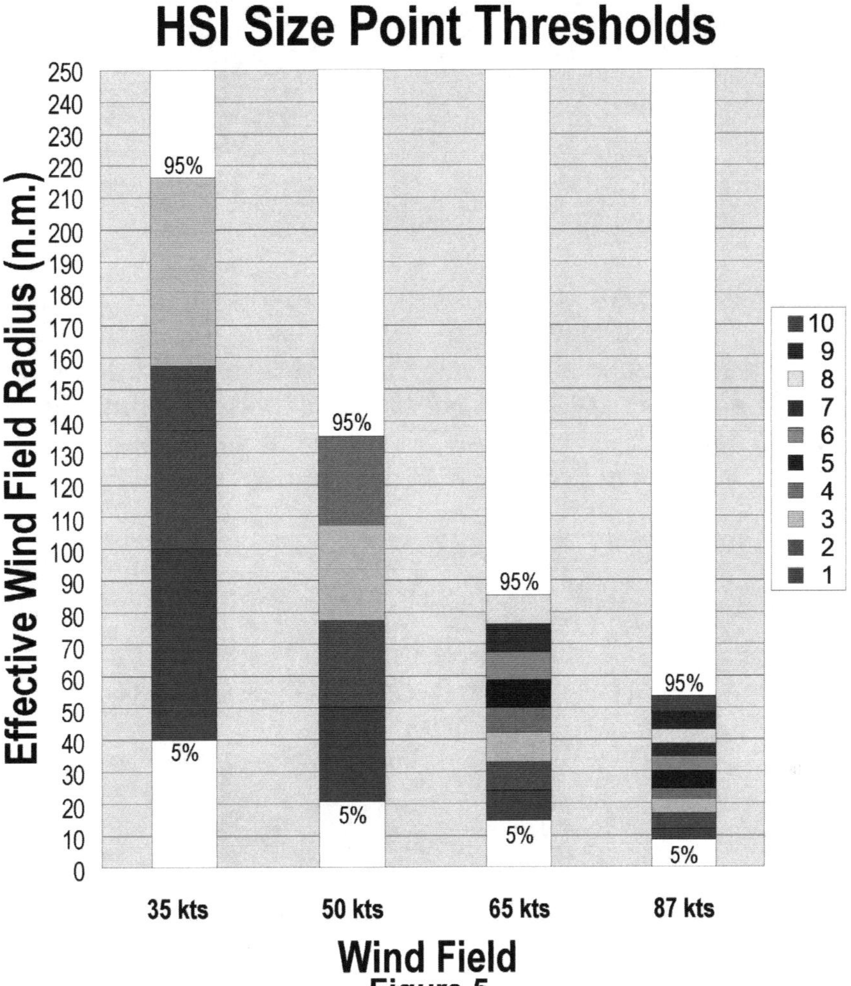


**Figure 5**

Points are awarded for each of the four wind radii depending on the extent of each wind field. For example, if a hurricane with 100-knot (115-mph) winds has a 35-knot (39-mph) effective wind radius that is in the top 1/3 of the 35-knot range, it would get three points for its 35-knot wind field. If its 50-knot (58-mph) effective wind radius is in the lower ¼ of the 50-knot range, then it gets an additional one point. If its 65-knot (75-mph) effective wind radius is in the second 1/8 of the 65 knot range, then it gets an additional two points. Finally, if its 87-knot (100-mph) effective wind radius is in the third 1/10 of the 87-knot range, then it is assigned an additional three points. This 100-knot hurricane would receive a total of 3+1+2+3 = 9 size points. A complete breakdown of the possible points for any tropical cyclone from a tropical depression to a Category 5 hurricane is shown in **Table 2**.

| Saffir-Simpson Hurricane Scale vs. HSI | | | | | | |
|---|---|---|---|---|---|---|
| Saffir-Simpson Classification | HSI Size | | HSI Intensity | | Total HSI | |
| | Low | High | Low | High | Low | High |
| Depression | 0 | 0 | 1 | 1 | 1 | 1 |
| TS | 1 | 7 | 1 | 4 | 2 | 11 |
| Cat. 1 Hurricane | 3 | 15 | 5 | 7 | 8 | 22 |
| Cat. 2 Hurricane | 3 | 25 | 8 | 10 | 11 | 35 |
| Cat. 3 Hurricane | 4 | 25 | 11 | 13 | 15 | 38 |
| Cat. 4 Hurricane | 4 | 25 | 15 | 20 | 19 | 45 |
| Cat. 5 Hurricane | 4 | 25 | 22 | 25 | 26 | 50 |

**Table 2**

A good recent example of how the Hurricane Severity Index can be used as a tool for estimating the destructive potential of a land-falling tropical cyclone would be a comparison of Hurricanes Dennis and Ivan. Both hurricanes struck the same area of the Gulf Coast. Each was a Saffir-Simpson Category 3 hurricane with 125-mph sustained winds at landfall. However, the resemblance ends there. **Figure 6** demonstrates the differences between the two hurricanes as they neared landfall.

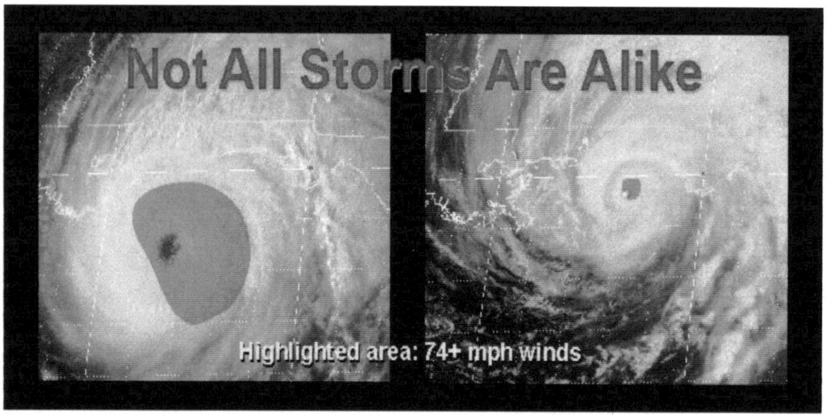


| Hurricane Ivan | | Hurricane Dennis | |
|---|---|---|---|
| Max Winds | 105 kts | Max Winds | 105 kts |
| Storm Surge | 10-15 ft | Storm Surge | 6-9 ft |
| Damage | $14.2 billion | Damage | $2.2 billion |
| Intensity Pts. | 12 | Intensity Pts. | 12 |
| Size Pts. | 21 | Size Pts. | 6 |
| Total HSI | 33 | Total HSI | 18 |

**Figure 6**

The red highlighted area in **Figure 6** represents the extent of the hurricane force or greater wind field in Hurricane Ivan on the left and Hurricane Dennis on the right. Even though Hurricane Ivan had weakened considerably by the time it neared the Mississippi Delta, its wind field was considerably larger than that of Hurricane Dennis. Hurricane Ivan's size point total (21) was 3.5 times that of Hurricane Dennis (6). This difference in the size of the wind fields of the two hurricanes resulted in vastly different effects both offshore and at landfall.

Hurricane Ivan produced significant wave heights of 50-55 feet across and maximum wave heights approaching 100 feet in the lease areas offshore southeast Louisiana. Hurricane Dennis produced significant wave heights of 30-35 feet and maximum wave heights of from 40-45 feet. Hurricane Ivan's storm surge approached 15 feet in Escambia Bay just east of Pensacola, FL. Hurricane Dennis produced a storm surge of between 5-7 feet just east of where the center made landfall between Pensacola and Panama City, FL. Hurricane Ivan produced damage totaling over $14 billion dollars in insured and uninsured losses. The losses from Hurricane Dennis were estimated to be near $2.2 billion dollars.

From a comparison of the damage produced by these two hurricanes with identical maximum sustained winds, it is obvious that the Saffir-Simpson Hurricane Scale is deficient in that it does not differentiate between two hurricanes of identical intensity but vastly different wind field size.

**Incorporating Hurricane Timeline Tools Into Your Hurricane Action Plan**

Now that each of these Timeline Tools has been discussed, we can examine how they might be incorporated into a hurricane action plan. A typical hurricane action plan involves a phased approach. In general, earlier phases involve tasks that would be less costly in terms of dollars as well as personnel time. Subsequent phases involve actions that are more costly, up to a complete shutdown of operations and/or relocation at a backup facility. An example of such a plan can be seen in **Table 3**.

| Phase | Actions Taken |
|---|---|
| A | Notify team members of potential threat. Check status of critical personnel. Book a percentage of hotel rooms. Check supplies and emergency equipment. Check on helicopter availability for evacuations. |
| B | Release ride-out and remote operations teams for personal preparations. Arrange for equipment rentals. Begin staging of critical equipment and supplies. Schedule evacuation of non-critical personnel. |
| C | Initial deployment of team to set up remote operations center. Book remaining hotel rooms. Ride-out team return to continue preparations. Communicate potential office closing time to employees. Evacuate non-critical personnel. |
| D | Full remote team and ride-out team in place. Announce office closing. Prepare for final shutdown of facility. Evacuate all remaining personnel. |
| E | Switch operations to remote site. Perform full shutdown of facility. |

**Table 3**

Assuming that your company has developed such a phased plan, certain trigger points have been identified for activating each phase. It's important to identify each trigger point as either hard or soft. A soft trigger point involves actions that have a somewhat flexible timeline. For example, checking on supplies when a distant potential threat is identified, or reserving hotel rooms for employees that might have to evacuate to a backup facility. These actions need to be taken at some point, but there isn't a critical timeline when they *must* be done. There is some flexibility in when these soft trigger actions are done. In contrast, a hard trigger is one that has a rigid timeline. If it takes a company a full 72 hours to safely shut down its facility (perhaps an oil refinery), then a decision to shut down or to not shut down has to be made prior to that 72-hour point before possible impact.

Assuming that a company has identified these soft and hard triggers and has developed a specific timeline for each action that must be completed prior to a possible hurricane impact, the next step is to define what is used to activate each trigger. A company might use subjective guidelines for escalating its hurricane plan from one phase to another. Subjective guidelines might be the distance to a developing tropical cyclone or the movement of a tropical cyclone within a certain area. This is where most plans fail. The use of subjective guidelines leads to an inconsistent and sometimes unnecessary activation of a hurricane action plan.

The Hurricane Timeline Tools can be used to provide objective guidance in the event of a hurricane threat. These tools take into account the latest 5-day forecast of a tropical cyclone's track, intensity and size. Average forecast error is also taken into consideration. For most plans, objective guidance can be derived from the following:

- 58-mph Probability of Wind Impact
- Worst-Case Scenario Arrival time of 39-mph Winds
- Forecast Arrival Time of 39-mph Winds
- Current and Projected Hurricane Severity Index

The key to using the 58-mph PWI for the activation of a hurricane action plan lies in understanding what probabilities are likely to exist at any given time up to five days before possible impact. Hurricane Rita from 2005 can be used as an example. Rita works very well as far as its track because it threatened a large part of the western Gulf of Mexico at one time. But just using the wind probabilities that existed for Rita doesn't address the issue of hurricanes much larger or smaller than Rita. Because the PWI depends greatly upon the size of the wind field. A larger hurricane would result in 58 mph PWI values much higher than for a smaller hurricane. **Figure 7** represents the range of probabilities that would typically exist from five days before landfall for hurricanes of various sizes.

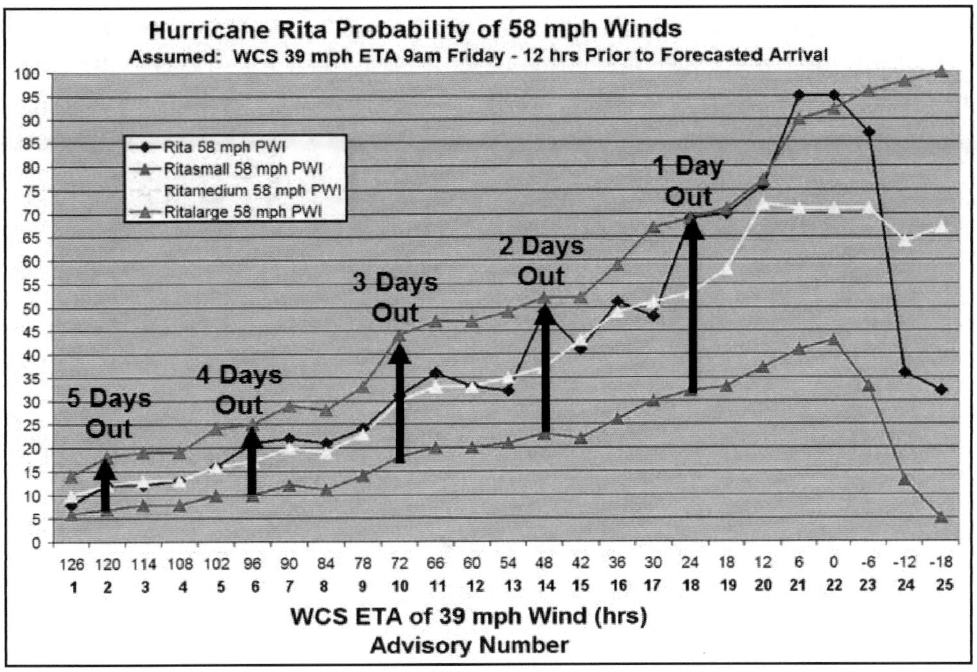


**Figure 7**

In **Figure 7**, the vertical scale represents the 58 mph PWI. The horizontal scale represents the shortest possible time before the arrival (in a worst-case scenario) of the tropical storm force (39-mph) wind field. The blue plot represents the actual 58-mph PWI for Rita. The green plot uses Rita's forecast track but with a 58-mph wind radius that is much smaller than average. The yellow plot assumes an average-sized Rita. The red plot represents a very large Hurricane Rita. From the graphic, we can derive a range of probabilities that exists for various-sized hurricanes threatening a location from five days before landfall (as shown in **Table 4**).

| | 5 Days Out | 4 Days Out | 3 Days Out | 2 Days Out | 1 Day Out |
|---|---|---|---|---|---|
| **Range** | **6-18%** | **10-25%** | **17-44%** | **22-52%** | **31-69%** |
| Avg. | 12% | 17% | 30% | 37% | 53% |

**Table 4**

The numbers in **Table 4** represent the ranges and the averages of the 58-mph PWI for a hurricane using Rita's forecast track but with small, average, and large-sized 58-mph wind fields. From the table, we can see that when a hurricane is five days from impacting a location, the range of 58-mph PWI values is from 6% to 18%, yielding an average value of 12%. Four days before the earliest possible arrival of tropical storm force winds, the average 58-mph PWI is in the 10% to 25% range, or about 17% on average. This probability increases rapidly within four days from impact for a hurricane moving toward a particular location.

Knowing these 58-mph PWI values and the earliest possible arrival time of tropical storm force winds (Worst-Case Scenario), a company can integrate objective guidance into its hurricane action plan. A sample plan can be seen in **Table 5**.

| **Potential Threat** | **HRI+ = Hurricane Threat Identified** | | |
|---|---|---|---|
| **Probable Threat Phase A** | 58 mph PWI > 15% | 39 mph WCS <96 hours | **HSI Decision Aid:** |
| **Probable Threat Phase B** | 58 mph PWI > 30% | 39 mph WCS <72 hours | Max forecast **HSI > 25** |
| **Probable Threat Phase C** | 58 mph PWI > 35% | 39 mph WCS <60 hours | Accelerate all actions |
| **Imminent Threat** | Forecast of TS force wind in <36 hours | | |

**Table 5**

The first tool used in the activation of a hurricane involves the early identification of a possible hurricane threat. This is the Hurricane Risk Indicator. A hurricane risk may be identified days before the National Hurricane Center begins to track a developing storm. At this stage of the plan, a company could begin some of the early preparations. Generally, this would involve steps that are fairly easy to do and don't cost much. The two main tools that are monitored next would be the 58-mph PWI and the worst-case arrival time of 39-mph winds. Depending upon a company's risk tolerance and its action timeline, it can design its plan around the 58-mph PWI values that typically exist at each time period prior to possible impact. Using the worst-case arrival time of 39-mph wind assures that the company takes into consideration forecast track error. A hurricane could move more quickly than forecast, arriving from 24-36 hours earlier than predicted.

In the example in **Table 5**, Phases A, B and C are activated only when the objective guidance reaches set critical values. But there are times when it might be prudent to activate the next phase of a plan a little early. Perhaps a company is sitting at "Phase A", waiting for the 58-mph PWI to reach 30% and the Worst-Case Scenario arrival of 39-mph wind to drop to 72 hours or less. Let's say it's 3pm on Friday afternoon and the 58-mph PWI is 28% (and rising) and the 39-mph WCS arrival time is 74 hours. The two conditions for escalating to Phase B are not quite there, but they're close. In this case, a company can use the projected Hurricane Severity Index to advance to Phase B a little early. In the example above, this early activation of a phase is allowed if the Hurricane Severity Index is projected to be 25 or higher. Such large and powerful hurricanes typically cause a company's competitors (and the general public) to react earlier than for smaller hurricanes. So an early escalation to Phase B is warranted.

Within 48 hours of landfall, forecast error drops significantly enough to rely more on the forecast arrival time of 39 mph winds rather than the Worst-Case Scenario arrival time. By this time, either the hurricane is moving toward a location or it isn't. The forecast arrival time of tropical storm force winds can be used for that final critical phase escalation which might involve a total shutdown of the facility. Using the forecast arrival time of tropical storm force winds takes into account the size of a hurricane's wind field as well as its speed of movement. This is much better than relying on using distance to the center of a hurricane as guidance.

## Summary

Hurricane Timeline Tools can allow a company to design and implement its hurricane action plan to take into account the level of risk it's willing to accept. Clear, objective guidance at each step of the way can allow risk managers to focus on the tasks involved in the activation of the hurricane plan rather than spending their time discussing whether or not it is time to advance from one phase of the plan to another. Objective guidance can help to eliminate the costly escalation of a hurricane action plan too early. More importantly, objective guidance can help to prevent the even more costly failure to activate critical phases of the hurricane action plan early enough to provide for a safe shutdown of operations and quick recovery after the hurricane threat passes.

*Authors' address: ImpactWeather, Inc,*
*P.O. Box 751869, Houston, TX 77275*
*email: Chris Hebert chebert@impactweather.com, Bob Weinzapfel rweinzapfel@impactweather.com*

## 146b Performance of and Mitigation Options for Vulnerable Plant Components

**James R. (Bob) Bailey**, Exponent, Inc., 10855 Richmond Avenue, Suite 175, Houston, TX 77042

Process units that reside within a petrochemical plant consist of multiple components, from process towers and pipe racks to furnaces and flare towers. Each component will have varying degrees of vulnerabilities to extreme winds and to storm surge associated with a hurricane striking the plant site. Among the more vulnerable components within a petrochemical plant are cooling towers, tanks, and electrical systems, specifically utility poles, transformers, and switchgear. Damage to these components can cause prolong disruptions to plant operations, resulting in high business interruption costs. This presentation will address the nature of hurricane damages sustained by cooling towers, tanks, and electrical systems. Also discussed are ways to quantify risks to these plant components in a manner that will enable operators and owners to make informed, cost-effective decisions regarding mitigation planning and associated reductions in loss exposures. Such knowledge is important from a commercial insurance perspective, and also from the standpoint of increasing operational reliability while decreasing potential business interruption losses.

# IMPROVING CHECK LISTING WITH PREDICTIVE FAILURE ANALYSIS©

*Jack Hipple*
*Innovation-TRIZ*
*Tampa, FL*

*Biography*

*Jack Hipple is Principal with Innovation-TRIZ, a consulting firm focused on organizational problem solving and specializing in the use of the breakthrough problem solving process known as TRIZ ("Theory of Inventive Problem Solving"). He graduated from Carnegie Mellon University with a BSChE and spent 30 years in the chemical industry prior to starting his own consulting business in 1999. His industrial career included responsibility for Dow Chemical's Discovery Research program and its global chemical engineering laboratory, project management responsibility for the National Center for Manufacturing Sciences, and new product development responsibilities for Ansell Edmont and Cabot Corporation*

## Abstract

Check listing is a very common way of making sure that we have thought about everything of concern to us, whether it be going to the grocery store, getting ready for a presentation, making sure that all our instrumentation is working in a chemical plant, or ensuring that our evacuation procedures are up to date. In the chemical industry, we primarily use HAZOP as our mechanism for doing this. It consists of asking basic questions about our process as designed and operated, questioning flow, temperature, pressure, and compositional variables. In the chemical industry we see many accidents that occur despite these reviews. In many of these cases, process review results were not implemented or changes were made in violation of known procedures, but in other cases, we find the root cause after the fact. We add this learning prior to our next designs. The nature of the questions we ask in these reviews is "what?" They basically do not question the basic design proposal or operation unless a serious and obvious flaw is identified.

How can we improve our ability to ensure that accidents don't happen? By becoming deliberate saboteurs and thinking about <u>*how*</u> to cause a problem, and not thinking in a traditional way about <u>*what*</u> can cause the problem. In a traditional HAZOP review, much caffeine is consumed going through every process variable and asking questions like, "higher, lower, none, combined with, etc". This is a tedious process with little energy. However, if we ask the participants to cause the process to fail, produce hazardous releases, injure people, etc. the mental process changes in a very sadistic, but productive way. When asked to figure out how to make a process release hazardous materials <u>*all*</u> the time and cause loss of human life, and an environmental catastrophe, a group will not only find many more possibilities, but will have fun doing it. This presentation will review several major incidents from this perspective and demonstrate how this technique could have been used proactively to improve the process design and minimize the potential for the disasters that happened.

## Basis for Approach

In the 1950's an insightful Russian inventor and patent examiner named Genrikh Altshuller determined that, after analyzing hundreds of thousands of breakthrough patents, that there was a basic inventive algorithm that could solve virtually any engineering problem. He also found there were similar and a limited number of generic inventive principles used across all areas of technology. This work led to "TRIZ" (Russian acronym for "Theory of Solving Inventive Problems"). Altshuller and his colleagues continued their work, continuing to develop these algorithms through the 60's and 70's. TRIZ emigrated from the former Soviet Union to the United States, Europe, and Japan in the late 1980's after Perestroika and is now in use by many Fortune 500 companies to solve difficult problems, improve intellectual property filings, and accelerate the pace of internal innovation. Major users include Motorola, Dow Chemical, Unilever, Siemens, Intel, the US Navy, British Petroleum, and Hewlett Packard. AIChE and ASME currently offer public courses in this methodology. The technology has been extended in its application to not only technology applications, but to business problems, human factors and ergonomics, biological systems, and intellectual property filing and circumvention. It is also a course offering by AIChE and ASME[5]

## The TRIZ Algorithm

In analyzing the global patent literature and its few (less than 5%) truly breakthrough patents, a basic algorithm for problem analysis was developed and can be presented in this simplified manner:

1. State the Ideal Final Result for the system, process, or component of concern. An example might be that a chemical process controller adjusts its settings automatically without human intervention for ANY process condition. This type of problem definition is very difficult for experienced engineers to do as it assumes a state that may be difficult to envision.

2. What resources do we have available to accomplish this? Are they sufficient to achieve this state? TRIZ looks at resources in a very broad sense and includes not only materials and substances, but time (before and after), space (within, around), fields (ex: thermal, chemical, electronic, magnetic), information (both collected and not collected, information being generated but user unaware of), and functional (conversion of one field or effect to another, i.e. mechanical to thermal via friction).

3. If steps 1 and 2 are insufficient, we usually find a contradiction that prevents achieving this ideal state. TRIZ uses a number of tools to resolve contradictions including the TRIZ 40 Principles and Contradiction table, as well as TRIZ separation principles for physical contradictions.

4. In addition to problem solving, the analysis of breakthrough inventions also revealed a series of lines and patterns evolution that could be used to forecast and plan technological and business developments.

This algorithm has been further developed and fine tuned over time, as has happened with any field of science, and has been labeled according to the year of its development. ARIZ-85 ("Algorithm of Inventive Problem Solving" from 1985) is the most commonly used version at present and has a total of 16 steps (vs. the simplified few stated previously), many of which assist the problem owner in defining the problem more precisely.

## TRIZ in "reverse": Predictive Failure Analysis©

Around 1970, some of Altshuller's students took the basic TRIZ problem solving algorithm and inverted it for use in analyzing and predicting failures. Its first use was in a political system problem in the former Soviet Union. The simplified algorithm, in reverse, is as follows:

1. State the Ideal Final Result (ex: We want the chemical process to operate safely at all times without injury to personnel or release of chemicals to the environment).

2. Invert this ideal result (ex: We want the chemical process to operate unsafely at times and occasionally cause injuries and release of hazardous substances)

3. Now comes the key step-----exaggerating this inverted ideality statement. (Ex: We don't want this process to *ever* run safely. We want *constant* release of hazardous materials resulting in *mass* injury within the external community, *large* environmental damage, etc.). We want the plant manager to be thrown in jail and the corporation sued for millions of dollars, threatening the survival of the corporation. This extreme exaggeration is the key to forcing participants to think about the absolute worst case and how to cause it. This is not only a key step, but the most difficult. It is analogous to the difficulty of defining an Ideal Final Result in normal TRIZ problem solving. It is important that a strong adjective ("always","never", "make sure", "completely") be part of this statement. If it does not contain such a strong word, it is unlikely that the statement is sufficient.

4. Now we ask, "How would we accomplish this?" "What resources do we need?" "How could we convert existing resources into a negative resource?" In this step, we utilize the comprehensive list of resources developed for TRIZ problem solving: Time, space, fields, field conversion, materials, and information.

Experiences with this approach have shown that it can improve root cause analysis as well as do a better job of preventing accidents and safety incidents. It has been used, not only in the chemical industry, but in banking transactions, software reliability, and food safety. This inverted TRIZ algorithm has been also referred to as Anticipatory Failure Determination©.

## Examples

Assuming that we have successfully completed the first three steps of this analysis (ideal result, inverted ideal result, exaggerated inverted ideal result), we look at time, space, fields, field conversion, substances and materials, time, and information as resources that can be used to accomplish this extreme negative situation. Let's look at an example of each.

Time as a Resource for Failure

The CCPS Process Safety Beacon in CEP Magazine (January, 2006) presented a summary entitled "Time Sensitive Chemicals", highlighting the fact that many chemicals have shelf life limits and become unstable or reactive with time in storage. Inhibitors in monomer storage systems have time limits as they are consumed. In principal, this is no different than freshness dates for food in a grocery store. In the sense of this methodology, we would ask, "How much time do we need to allow decomposition? How would I MAKE SURE that this amount of time is available?" The answer may lead us to look closer at process storage, inventory, shipping delays, as well as customer inventory time. Some of these segments may be seen by a checklist review, but if we focus on HOW to create the time, we will tend to see more possibilities, such as changes in the customer's inventory (not just our own). We would more appropriately consider the relationship of time and temperature. Slow rail deliveries in an excessively hot summer may result in consumption of an inhibitor. This has been the case in styrene polymerizations within rail cars.

In a previous example presented at the 2007 CCPS meeting[4], the rupture of a hazardous chemical drum shipment was traced back to a long overseas transport time, allowing a corrosive reaction (iron plus bromine) sufficient time to generate enough heat and pressure to go beyond the pressure rating of the drum. The combination of time, extremely low heat capacity (0.1 vs. water's 1.0), and the heat of corrosion contributed to a dangerous release of bromine.

In an ethylene oxide explosion[3], long standing rust (a polymerization catalyst) in a dead spot (*time!)* had accumulated. How can we *make sure* that we can accumulate reactive materials? How would we design a piping system to do this? Then don't do it!

We frequently recognize time as a potential problem area, but we don't often exaggerate its impact as we do with Predictive Failure Analysis©.

Materials as Resources for Failure

We are generally aware of the potential consequences of mixing the wrong materials together, but we sometimes forget that normally "non-hazardous" materials can create danger, if combined in the wrong way. We also recognize in general the hazards of confined spaces. The combination can be deadly. CCPS reported this past August on the death of a worker using a sheet of black plastic to shield a black light to inspect piping. 150 feet away an open nitrogen line put enough nitrogen into this "closed" space, causing asphyxiation. The checklist approach did not consider checking this possibility. Suppose we had used this type of analysis in this case. We would ask the question, "How can be *make sure* there is no breathable air within this space? Are there any materials nearby that might allow this to happen? How would we *make sure* that nitrogen was introduced into this space? Where is there nitrogen?" The source would probably have been identified and eliminated.

In the previously mentioned bromine release case, the material being shipped and the material from which the drum was constructed were material resources for a runaway reaction beyond the obvious corrosion reaction concern.

In another well publicized case, an explosion within a kerosene tank containing a water layer was traced to bacterial action generating methane which was the source, combined with a welding torch, for an explosion. The checklists used did not catch this, but if we had asked the question, "How would I make sure that a flammable gas was generated in the tank? What would be required? Is it possible to generate the materials?"

Substituting steel and titanium in dry and wet chlorine service is a frequent materials mistake.

Information as a Resource for Failure

Since virtually all of our processes measure and respond to information generated by a variety of sensors and instrumentation, it is important to consider there normally "good" resources and how they can be used in a negative way. In the now famous Texas City fire and explosion, a column level indicator showed level decreasing at the same time the column was overflowing. This occurred because the instrument was submerged and under those conditions, density was being measured, not level. As the temperature rose during column heat up and recirculation, the level "indication" (not the actual level) actually decreased whereas the column was actually overflowing. A back up level indicator failed to respond. Consider how things might be different, if during a safety review we had used this Predictive Failure Analysis© process instead of the normal HAZOP approach. We would have asked, "How could I *make sure* that we were *never* aware of that the level was? That the level indication was *never* correct?" It is probable that someone, during the review, would have suggested that, if the fluid level in the column was higher than the maximum level indication, the level indication would be meaningless and potentially dangerous. It might also have been suggested to find a mechanism to determine the status of any backup instrumentation.

The general question to ask is "How Would I *make sure* that my process instrumentation readings are incorrect *at all times?"*

Unavailable or unknown information causes the same potential for disaster. If explosion potential or explosive ranges are simply not known, the most well intentioned HAZOP team will not take precautions.[3]

Energy as a Resource for Failure

This is an area which we normally consider. We usually think about obvious and visible energy sources such as furnaces, steam, heat transfer fluids, hot water, gas, and electrical sources. In some cases, however, we miss the energy produced by internal processes and the accumulation of energy in ways and places not expected. In a 2005 acetylene cylinder facility explosion in Perth Amboy at Acetylene Services[5], a failed check valve allowed a flow of acetylene backward from a calcium carbide/water reactor into the water supply line and ultimately into a closed space which contained a space heater. Now consider if we had asked, "How can we *make sure* that acetylene flows backwards? How could we make sure that, if this happened, it would concentrate in a confined space? How would I *make sure* that, if this happened, there would be a source of ignition? Why not put a space heater in the space? This was well intentioned to keep operating personal warm in a water area that was never expected to contain acetylene.

The approximately 200 dust explosions in the last decade in the US have been analyzed by the Chemical Safety Board[5] and two key aspects have been pointed out. First, in addition to the traditional fire triangle, it is required to have dust concentrations be elevated and in a confined space. These are mechanisms for concentrating the energy of dusts. Then something disturbs the settled dust is dispersed and ignites. In analyzing our processes, we can ask the simple question (in addition to all the traditional ones), "How can we MAKE SURE that dust concentrates?" Inadequate ventilation? Cleaning? How can we MAKE SURE that we have a confined space for dusts to accumulate? How can we MAKE SURE we have both? How can we *make sure* that there is *always* an ignition source present?

In the 2005 Praxair gas cylinder storage explosion, energy sources came from the sun (it was an unusually hot 97 degrees in late spring) and hot asphalt, the combination of which was sufficient to overwhelm the pressure relief valves in the cylinders. We often think about the sun and asphalt as things that are just "there"—part of the facilities and background, but from a TRIZ analysis perspective, they are resources, and they can be energy sources. If we had asked the question, "How can we *make sure* that there is sufficient heat to heat the cylinders sufficiently to release gas?" it is possible that we might have thought about shielding the storage area or painting the surface with a reflective coating. We might have calculated more thoroughly the heat transfer and relief valve limitations of the system.

Fields as a Source of Failure

Basic fields include thermal, chemical, magnetic, electronic, acoustic, electromagnetic, and biological. The potential impact of recognized fields is normally considered in our check listing, but frequently field conversion is not (see next section). Unrecognized fields are the concern. The simple way to do this is to take this list and ask "is the field present?" If so, what are the consequences? Next, ask how could I generate such a field if it is not recognized as present already? Then, what are the consequences of this field's presence? We can use Table 1 as a template for doing this.

## Table 1

## Field Source Worksheet

| Field | Present? (Y/N) | If No, how to produce? | Potential Consequences |
|---|---|---|---|
| | | | |
| Mechanical | | | |
| Thermal | | | |
| Chemical | | | |
| Electronic | | | |
| Electromagnetic | | | |
| Acoustical | | | |
| Pressure | | | |
| Acoustical | | | |
| Human Behavior | | | |

Note that we have included "human behavior" as one of these fields. We have many automated chemical plants that we assume are not impacted by human behavior, but as we see frequently (and was the case in the PB fire and explosion) that humans can turn off alarms and ignore them. The question to be asked here is how could we *make sure* that an alarm or warning was ignored? How would we make that happen *all the time?* What resources are required? Are they present? Can they be created with the resources present? In the case of human behavior, we can think in terms of conflicting data presentations, parallel job responsibilities, personal experience and past history of false alarms, etc.

Forcing ourselves to create these various fields is a more comprehensive approach than simply asking ourselves whether they are present or not. Like an ignition source in a fire, we seldom know what it was, but it's always there. This is not to imply that all these fields and sources of energy are everywhere at all time, but to say that we frequently get blindsided by our narrow vision of the process environment.

Field Conversion as a Source of Failure

There is almost an endless list of field conversions. Chemical fields (reactions) generate heat, electrical fields generate both thermal fields and magnetic fields, and magnetic fields generate electrical fields. Chemical fields (reactions) are produced by electrical fields, both producing thermal fields. Mechanical fields (friction, rotating equipment, etc.) all produce friction, heat, and static. Static (an electrical field) is an ignition source for gases and liquid vapors. We know most of these facts. In the use of TRIZ in conventional problem solving, we usually are looking for field conversion in a positive sense, i.e. we are looking for potential resources for problem solving. In failure prediction, we are looking at the opposite: the potential consequences of these field conversions from a negative standpoint. What new fields can be formed and how could these new fields cause harm and damage?

As an example, the corrosion of materials is normally considered a nuisance or quality problem, and a factor that we use to choose materials of construction. However, corrosion is a chemical reaction and as such, has thermal effects. In the aforementioned bromine drum rupture, the concern was about the drums leaking a hazardous, corrosive material and causing personal harm and injury. The drums were properly designed (i.e. wall thickness) to withstand the corrosion rate over the intended time. However, they were not pressure vessels and were not designed to withstand the pressure generated from the liquid turning into a gas from the heat of corrosion. Aggravating this situation was the extraordinarily low heat capacity of liquid bromine (0.1 vs. water's 1.0). The combination of these two things in combination with the drums being overfilled, caused the hazardous has release. Again, if we had asked ourselves, "How could we *make sure* the drums ruptured in transit?" we would have said overfill them, let them sit in a hot Houston ship channel, and not properly calculate the heat balance on the drum during transport. In another case, corrosion of steel in an evaporator waiting to cool prior to vessel entry removed sufficient oxygen to produce a less than breathable atmosphere several hours after a vessel entry check had been performed, indicating safe entry. Several hours later, the maintenance worker entered the vessel without independent air supply and was immediately overcome and died before rescue was possible. This is an example of time, energy, and field conversion all contributing to a fatality.

In the BP raffinator tower explosion mentioned previously, the filled tower's thermal field applied unknowingly to recirculating fluid was changing the density of the fluid and affecting the mechanical force on the level indicator. Basic laws of physics tell us that no field can be used or generated without their being one or more accompanying fields. One of our responsibilities is to make sure that we are aware of all the potential byproduct fields a field, process, system, operation, or movement can produce and then methodically step through the consequences of these additional fields on our processes, equipment, instrumentation, and control. A simple table for doing this is shown in Table 2:

**Table 2**

**Field Conversion Worksheet**

| **Original Field** | **Byproduct Field 1** | **Byproduct Field 2** | **Potential Consequences** |
|---|---|---|---|
| | | | |
| Mechanical | | | |
| Thermal | | | |
| Chemical | | | |
| Electronic | | | |
| Electromagnetic | | | |
| Acoustical | | | |
| Pressure | | | |
| Optical | | | |
| Human Behavior | | | |

There are TRIZ (and other) handbooks[1,2] summarizing these field conversions, some of which are unknown in a practical everyday sense. Usually, in traditional TRIZ problem solving, we are trying to identify fields that can be cleverly used for solving a problem; in this case we are looking for by-product fields in a negative sense.

Some of these field conversion effects can be subtle and difficult to recognize and that is why it is critical to exaggerate our thinking in this process. Compressed air is frequently dried and purified through the use of molecular sieves. Nitrogen is preferentially absorbed and an enriched stream of oxygen is produced initially[3]. If an enriched oxygen stream is a hazard, this needs to be considered. This is a case of a chemical field (adsorption) producing a change in substance concentration, which can be considered as an enriched chemical field in terms of oxygen.

Note that human behavior has been included. This could include stress, long work hours, personal or family concerns, or supervisory/employee friction. The "byproducts" may not be traditional physical items, but certainly could include morale, attention span, and concentration. All of these have the potential to indirectly cause harm in a process through inattention, incorrect data collection and response, and incorrect decision making.

Space as a Resource for Failure

Insufficient space (confined space) is frequently cited as an accident cause. Using Predictive Failure Analysis© will not prevent doing things we already know are wrong or are not in compliance with known safety procedures, such as not using proper confined space procedures. However, space is a much broader term than this. It can be space around a system as well as space within a system (i.e. voids). It can be isolated spaces or dead spots in processing systems that can "hold up" materials for long periods of time (see previous section on time) and allow chemical reactions to proceed that are unanticipated.

A worker trying to clean tank residue was overcome by xylene fumes after a tank cleaning. Solids have voids (a space resource!) which can trap hazardous materials. Table 3 is an outline for use in evaluating space resources as resources for failures.

**Table 3**

**Space as a Hazardous Resource**

| **Type of Space Resource** | **Is This Space Available?** | **Potential Consequences** | **How to Create?** |
|---|---|---|---|
| | | | |
| On the surface | | | |
| External, surrounding space | | | |
| Space inside | | | |
| Porous space inside | | | |

## Conclusions and Summary

Checklists and diagrams are useful and necessary, but they can be incomplete and the process for completing them can be boring when what is needed for safety thinking is stimulation and completeness. Deliberately trying to make a process, system, or piece of equipment fail produces possibilities that check listing by whatever mechanism, does not.

## References


1. Mann, D. et.al. Hands On Systematic Innovation, Creax Press, 2003, pp.236-237, 345-354
2. Salamatov, Y. TRIZ: The Right Solution at the Right Time, Insytec, 1999, pp 245-253
3. Kletz, T. What Went Wrong?: Case Histories of Process Plant Disasters, 4th edition. Gulf Publishing, Houston, TX, 1998
4. Center for Chemical Process Safety Proceedings, 2007 www.aiche.org/ccps
5. Chemical Safety Board Safety Video, March 2007


## Resources


1. Mann, D. et.al. Matrix 2003, Creax Press, 2003
2. Hipple, Jack. "Using TRIZ in Reverse to Analyze Failures" Chemical Engineering Progress, May 2005, pp 48-51
3. Altshuller Institute for TRIZ Studies, Worcester, MA: www.aitriz.org
4. Innovation-TRIZ web site: www.innovation-triz.com
5. American Society of Mechanical Engineers web site: www.asme.org
6. On line TRIZ Journal: www.triz-journal.com

## 164a Potential Process Plant Impacts from Regulating CO2 as a Pollutant

**Susan Douglas**, Industrial Systems, CH2M HILL, Baton Rouge, LA 70801

This presentation will provide a high level overview of potential process impacts from anticipated regulatory changes associated with Climate Change legislation. Topics will include current information on the fallout from the Massachusetts v. EPA Supreme Court decision and the abridged version of Carbon Trading 101. Examples of what these changes could mean in terms of emission reductions within process units will be discussed, and select technologies currently under investigation will be reviewed.

# Economics of Plant Energy Saving Projects in a Changing Market

AIChE Spring
Meeting 2008
Douglas White
Emerson Process
Management

# Speaker

Doug White

Principal Consultant and Vice President, APC Services

Advanced Applied Technologies

Process Systems and Solutions

Emerson Process Management

Many years experience designing, justifying, installing and commissioning advanced real time computer applications in the process industries.

EMERSON Process Management

# Background

- Plant Energy Costs - often largest component of operating costs after feedstock – 8% of crude in ’96 to ’06 period

# Fuel Price Volatility

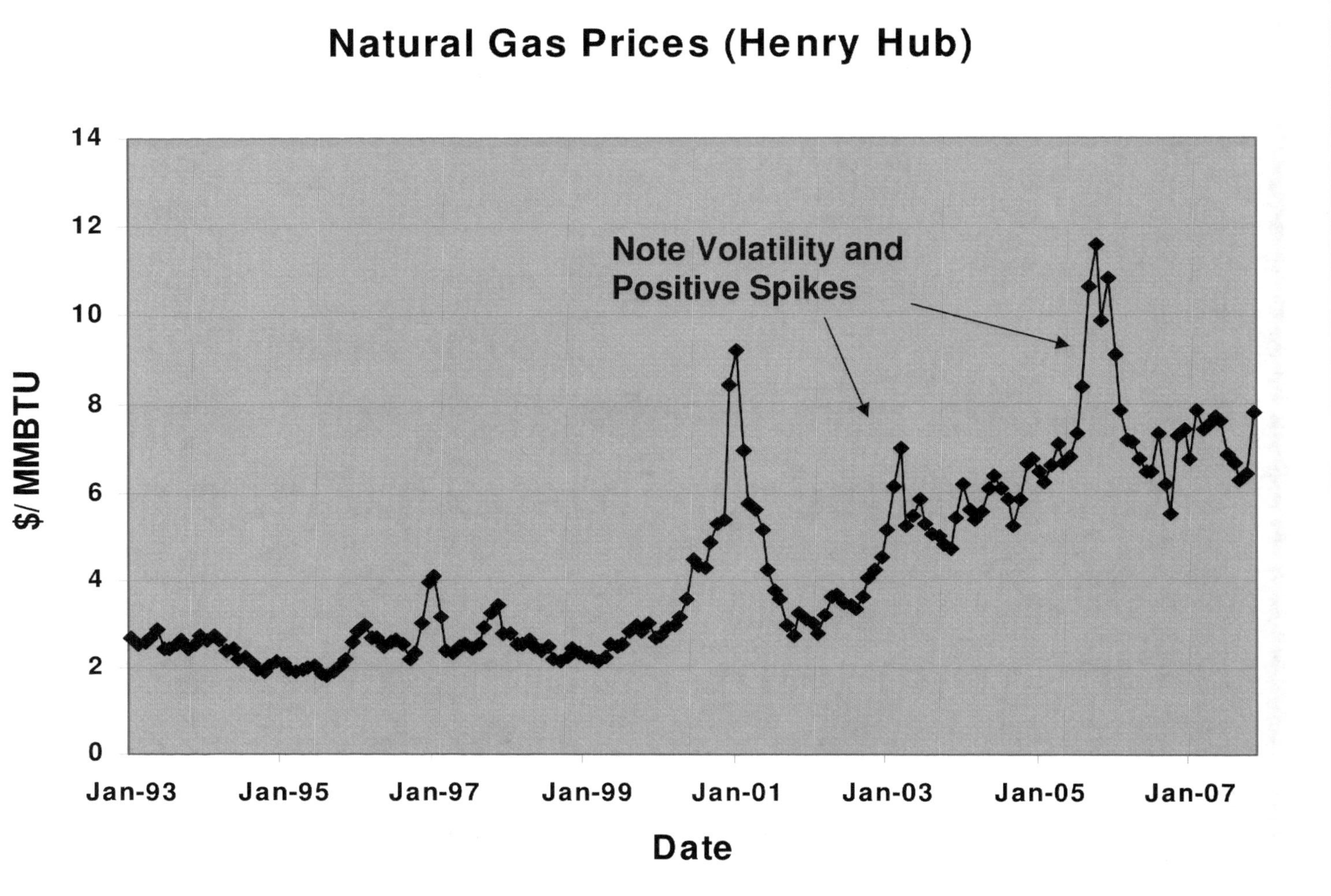


Source: US DOE EIA

# Questions

- How to analyze investments in energy projects when there is significant uncertainty on the pricing to be used for analysis
- How to analyze energy investments in flexibility – making equipment capable of using multiple fuels

# Value of Investment /Asset/ Plant

- → The financial value of any investment or asset equals the net present (discounted) value of the after tax "free cash flow" (ATCF) generated by the investment/ asset
- → The appropriate discount rate is a function of the corporations projected after tax weighted average cost of capital (WACC) with an adjustment for the "riskiness" of the investment compared with the normal "riskiness" of the business

# Standard Investment Analysis

$$NPV(ATCF) = \sum_{i=1}^{i=n} \frac{ATCF_i}{(1+r_c+r_r)^i} + \frac{CoV_{n+1}}{(1+r_c+r_r)^{n+1}}$$

$$NPV(IC) = \sum_{i=0}^{i=n} \frac{IC_i}{(1+r_c)^i}$$

If NPV(ATCF) greater than NPV(IC) the investment has a positive return

$ATCF_i$ = incremental after tax cash flow in period (year) i due to the investment in period (year) 0
$IC_i$ = investment sequence
n = number financial evaluation periods for the investment
$CoV_{n+1}$ = continuing value of the investment in period n+1
$r_c$ = corporate weighted average after tax cost of capital
$r_r$ = risk premium to be applied to the expected returns from a particular investment.

# Project Risks

- Internal
  - Demand
  - Technical – Performance
  - Project Execution
  - User Acceptance
- External
  - Price Risk
  - Financing Risk
  - Material and Labor Costs
  - Political and legal

# Handling Uncertainty – NPV Calculation

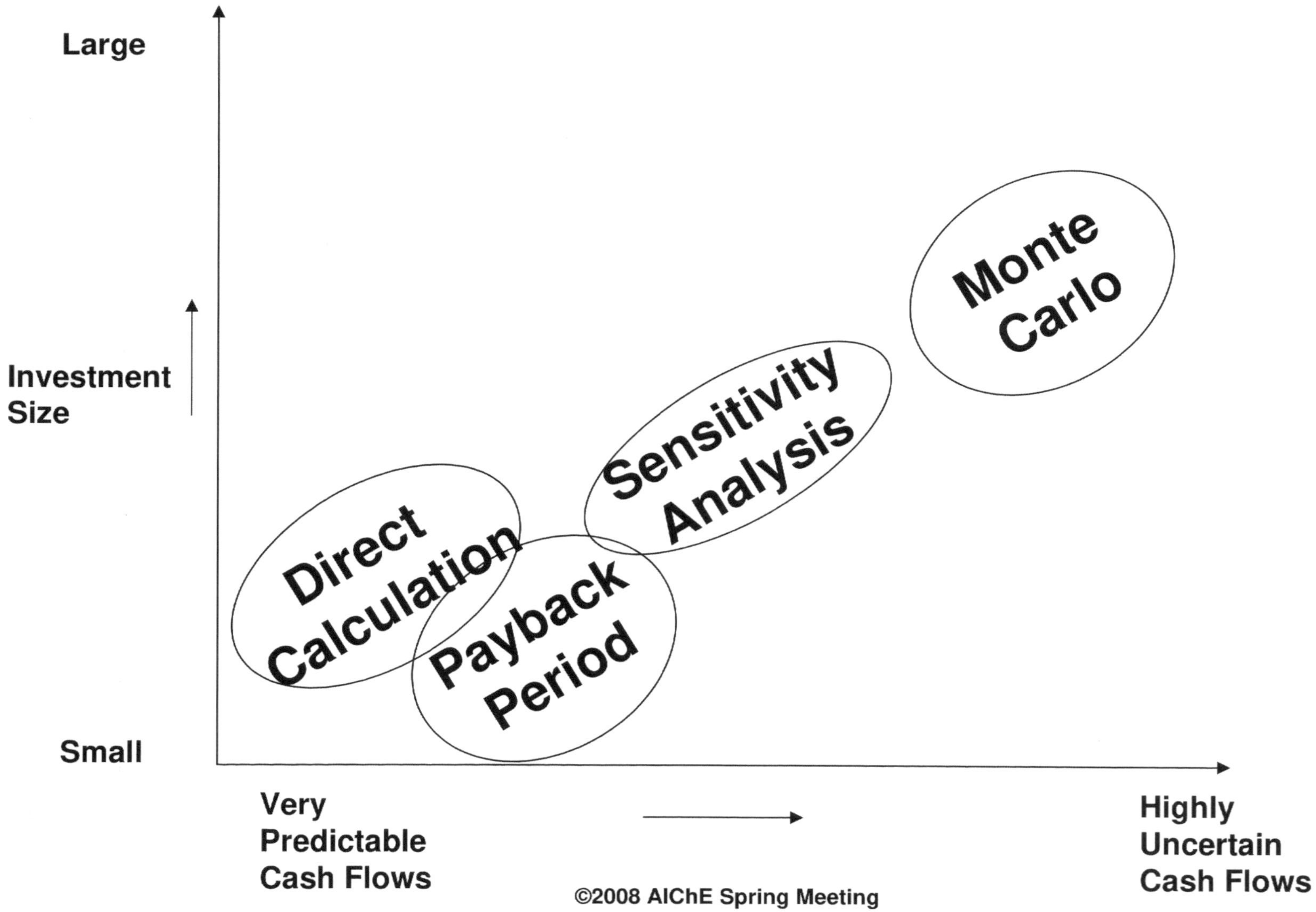

# Modeling Independent Variables – Normal Distribution

$$y(x) = \frac{1}{\sigma\sqrt{2\pi}} e^{-\frac{1}{2}\left(\frac{x - mean}{\sigma}\right)^2}$$

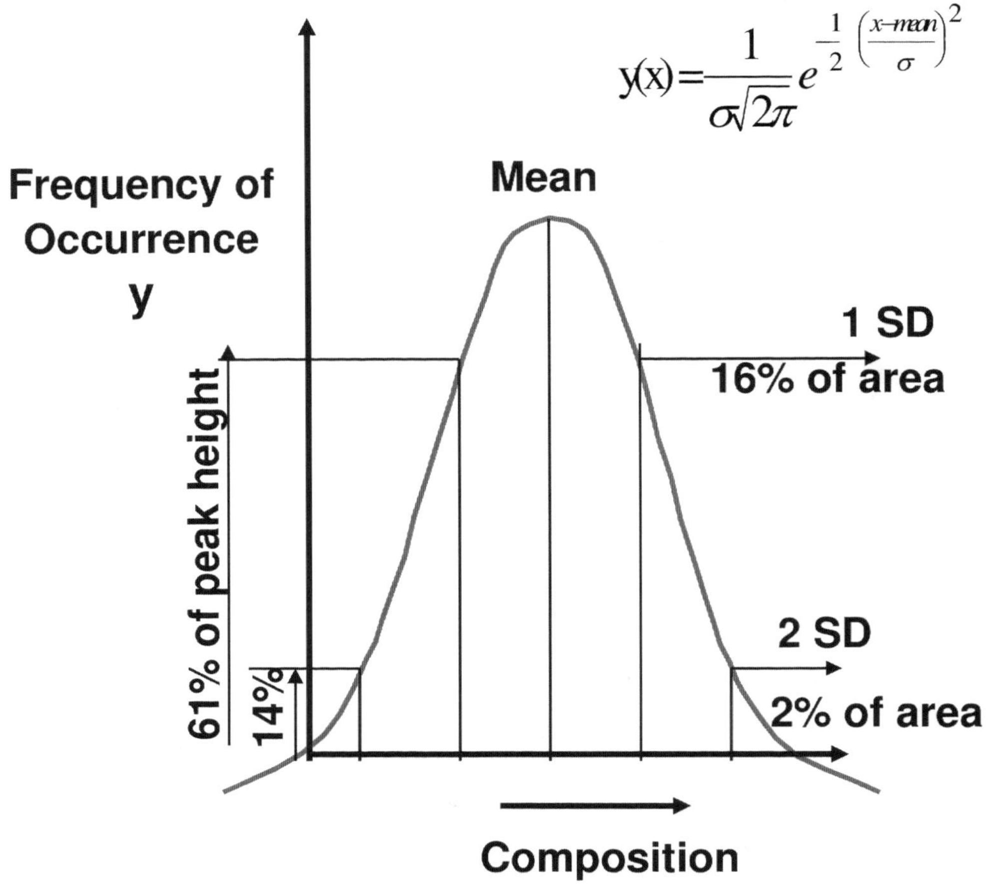

# *Representation of Uncertainty in Forecast*

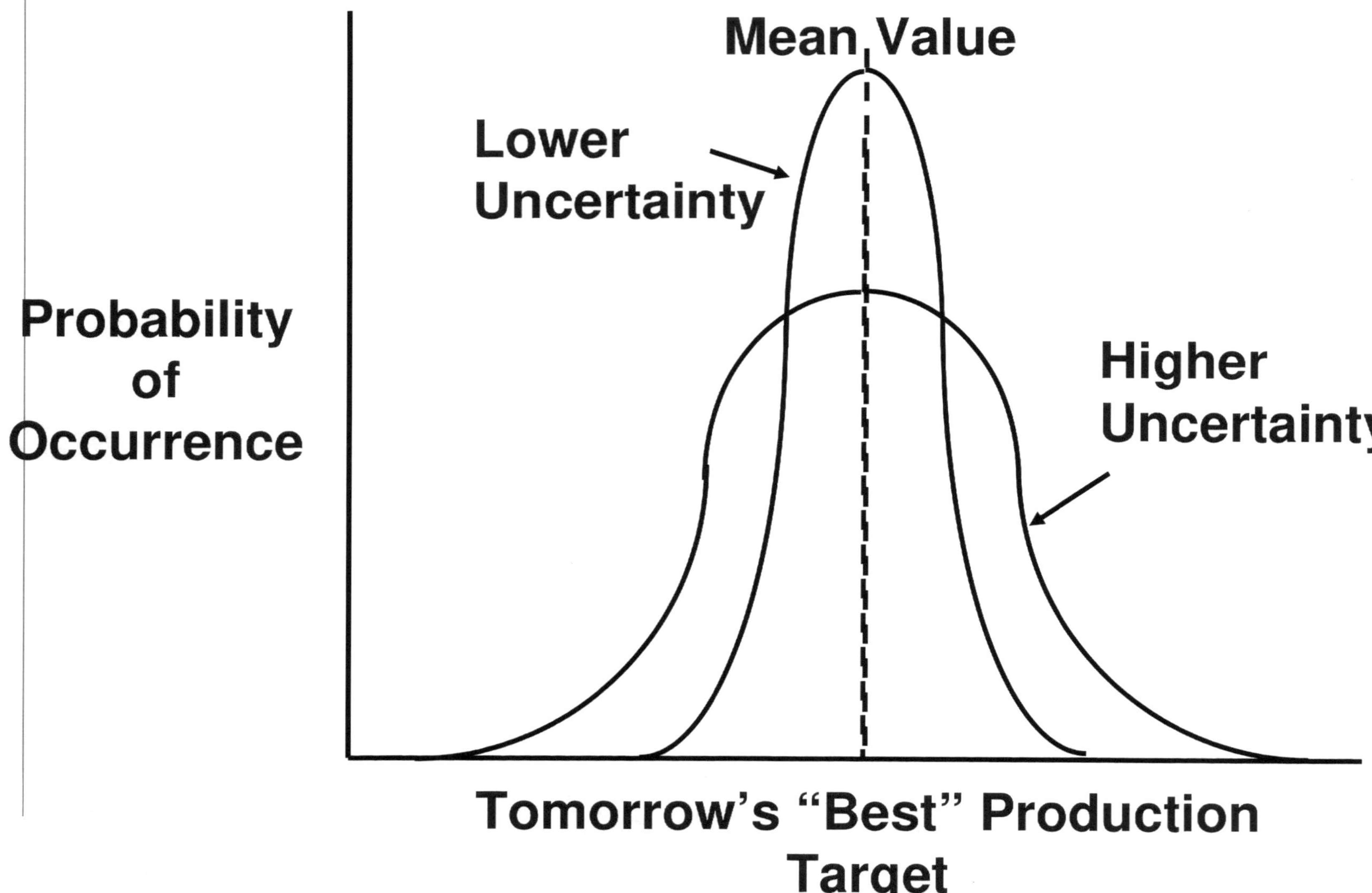

# Trend Line

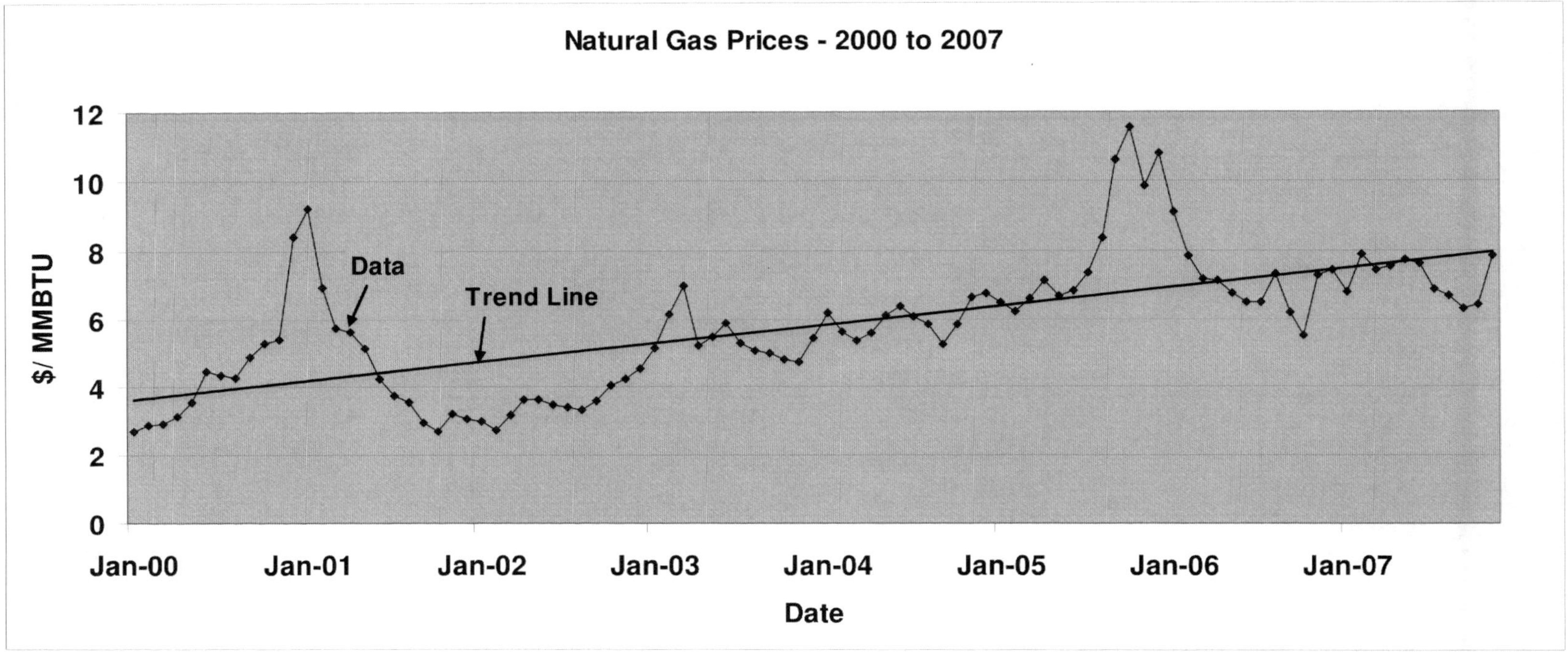

Natural Gas Prices - 2000 to 2007
12
10
8
6
4
2
0
$/ MMBTU
Data
Trend Line
Jan-00
Jan-01
Jan-02
Jan-03
Jan-04
Jan-05
Jan-06
Jan-07
Date

# Assumptions – Gas Trend Line

| Table 1 | | | | | |
|---|---|---|---|---|---|
| Assumptions | | | | | |
| Process Demand MMBTU/ Hr | | 150.0 | | | |
| New Heater Investment Cost | $ | 13,200,000 | Maintenance Costs Per Year, % Investment | 2% | |
| Initial Gas Price, $/ MMBTU | $ | 8 | Gas Price Increase Per Year, $ | 0.56 | |
| Old Equipment, Efficiency, % | | 75 | Equipment Life, Years | 10 | |
| New Equipment, Efficiency, % | | 90 | Operating Days/ Year | 355.0 | |
| Fixed Costs/ Yr | $ | 15,000 | Cost of Capital | 10% | |
| Depreciation/Yr | $ | 1,320,000 | Tax Rate | 34% | |

# Fixed Trend Line Calculational Inputs

Table 2 - NPV Base Case

| Year | 0 | 1 | 2 | 3 | 4 | 5 | 6 | 7 | 8 | 9 | 10 |
|---|---|---|---|---|---|---|---|---|---|---|---|
| Average Process Demand MMBTU/ Hr | | 150.0 | 150.0 | 150.0 | 150.0 | 150.0 | 150.0 | 150.0 | 150.0 | 150.0 | 150.0 |
| Fuel Cost, $/ MMBTU | | 8.00 | 8.56 | 9.12 | 9.68 | 10.24 | 10.80 | 11.36 | 11.92 | 12.48 | 13.04 |
| Fuel Cost Savings, $/ Yr | | 2,272,000 | 2,431,040 | 2,590,080 | 2,749,120 | 2,908,160 | 3,067,200 | 3,226,240 | 3,385,280 | 3,544,320 | 3,703,360 |
| Fixed Cost | | 15,000 | 15,000 | 15,000 | 15,000 | 15,000 | 15,000 | 15,000 | 15,000 | 15,000 | 15,000 |
| Maintenance Costs | | 264,000 | 264,000 | 264,000 | 264,000 | 264,000 | 264,000 | 264,000 | 264,000 | 264,000 | 264,000 |
| Depreciation | | 1,320,000 | 1,320,000 | 1,320,000 | 1,320,000 | 1,320,000 | 1,320,000 | 1,320,000 | 1,320,000 | 1,320,000 | 1,320,000 |
| Profit before Tax | | 673,000 | 832,040 | 991,080 | 1,150,120 | 1,309,160 | 1,468,200 | 1,627,240 | 1,786,280 | 1,945,320 | 2,104,360 |
| Tax | | 228,820 | 282,894 | 336,967 | 391,041 | 445,114 | 499,188 | 553,262 | 607,335 | 661,409 | 715,482 |
| Profit after Tax | | 444,180 | 549,146 | 654,113 | 759,079 | 864,046 | 969,012 | 1,073,978 | 1,178,945 | 1,283,911 | 1,388,878 |
| Net Cash Flow | (13,200,000) | 1,764,180 | 1,869,146 | 1,974,113 | 2,079,079 | 2,184,046 | 2,289,012 | 2,393,978 | 2,498,945 | 2,603,911 | 2,708,878 |
| *Net Present Value* | $42,944.17 | | | | | | | | | | |

## Used 10 Year Forecast for NPV Calculation

# Expected Investment Return – Fixed Trend Line

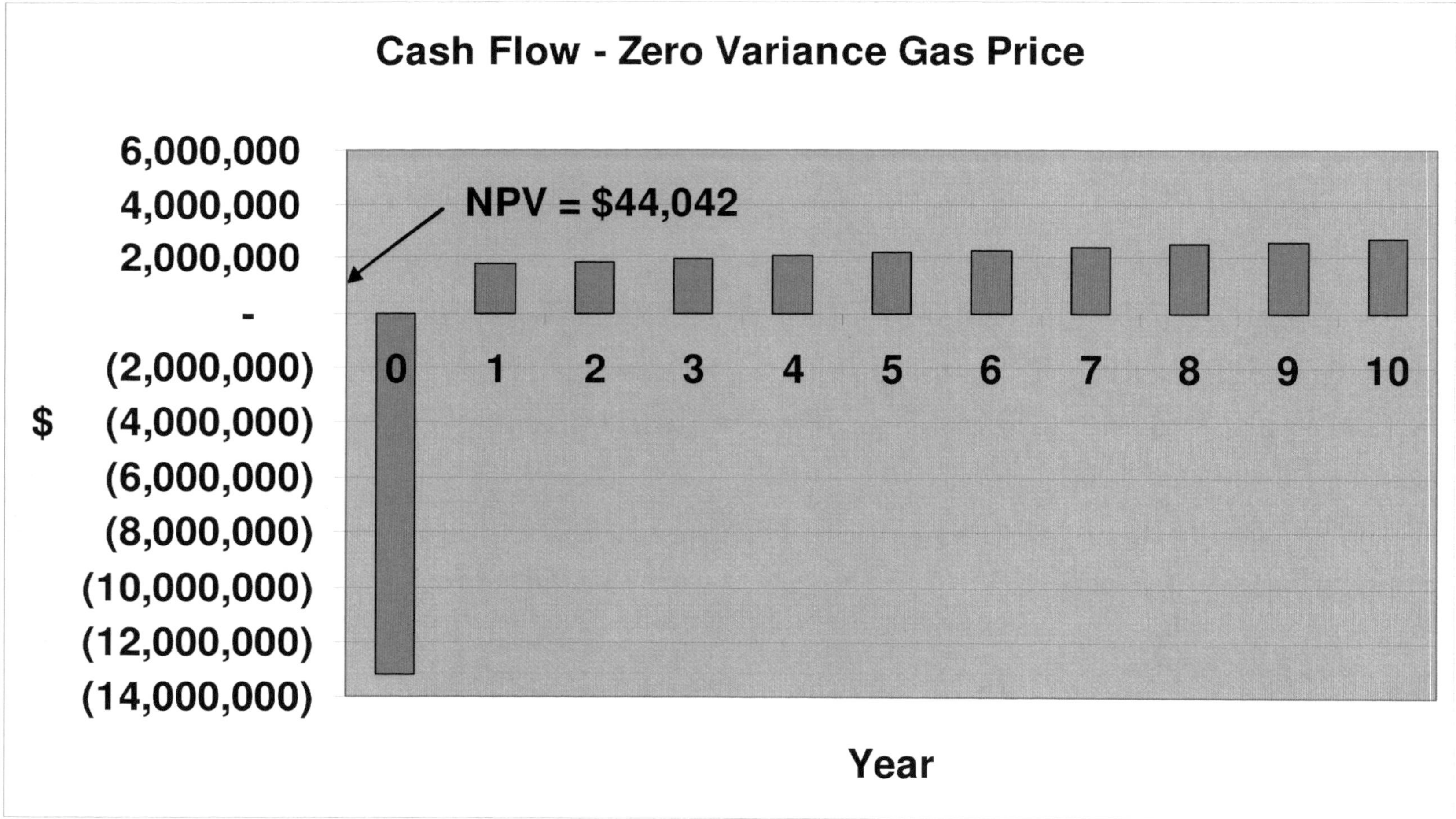

# Price Modeling

- Since prices for products cannot go below zero (or some equivalent lower bound) but are not bounded above, price distributions are typically unsymmetrical
- For this work, the prices were simulated with a Gumbel distribution,

# Trend Line Residuals Distribution

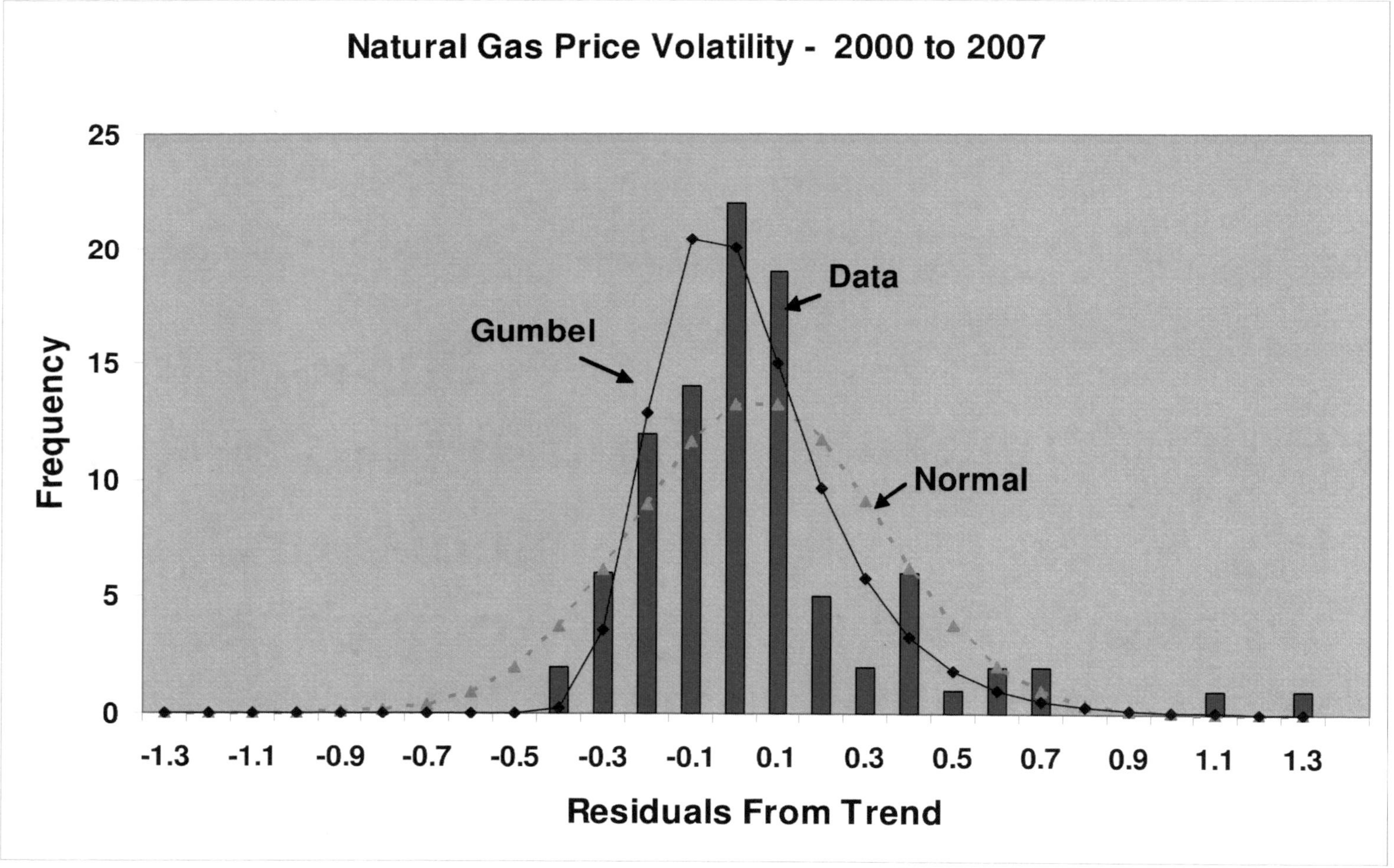

# *Monte Carlo Simulation*

- → Choose statistical distribution for independent variable(s)
- → Run large number of trials calculating the NPV with randomly chosen inputs for the independent variables fitting the chosen statistical distribution
- → Evaluate the expected value of NPV from the trials

# Varying Gas Price - Trials

| Table 1 | | | | | |
|---|---|---|---|---|---|
| Assumptions | | | | | |
| Process Demand MMBTU/ Hr | 150.0 | | | | |
| New Heater Investment Cost | $ 13,200,000 | Maintenance Costs Per Year, % Investment | 2% | | |
| Initial Fuel Price, $/ MMBTU | $ 8 | Fuel Price Increase Per Year, $/ MMBTU | 0.56 | Gumbel Variance | 0.0435 |
| Old Equipment, Efficiency, % | 75 | Equipment Life, Years | 10 | | |
| New Equipment, Efficiency, % | 90 | Operating Days/ Year | 355.0 | | |
| Fixed Costs/ Yr | $ 15,000 | Cost of Capital | 10% | | |
| Depreciation/Yr | $ 1,320,000 | Tax Rate | 34% | | |

# Single Trial Result

Table 2

| Year | 0 | 1 | 2 | 3 | 4 | 5 | 6 | 7 | 8 | 9 | 10 |
|---|---|---|---|---|---|---|---|---|---|---|---|
| Average Process Demand MMBTU/ Hr | | 150.0 | 150.0 | 150.0 | 150.0 | 150.0 | 150.0 | 150.0 | 150.0 | 150.0 | 150.0 |
| Fuel Gas Cost, $/ MMBTU | | 8.14 | 8.3889615 | 9.407838123 | 9.53127403 | 10.0810087 | 10.6417311 | 11.1167162 | 12.111305 | 12.573138 | 13.3433581 |
| Fuel Cost Savings, $/ Yr | | 2,312,340 | 2,382,465 | 2,671,826 | 2,706,882 | 2,863,006 | 3,022,252 | 3,157,147 | 3,439,611 | 3,570,771 | 3,789,514 |
| Fixed Cost | | 15,000 | 15,000 | 15,000 | 15,000 | 15,000 | 15,000 | 15,000 | 15,000 | 15,000 | 15,000 |
| Maintenance Costs | | 264,000 | 264,000 | 264,000 | 264,000 | 264,000 | 264,000 | 264,000 | 264,000 | 264,000 | 264,000 |
| Depreciation | | 1,320,000 | 1,320,000 | 1,320,000 | 1,320,000 | 1,320,000 | 1,320,000 | 1,320,000 | 1,320,000 | 1,320,000 | 1,320,000 |
| Profit before Tax | | 713,340 | 783,465 | 1,072,826 | 1,107,882 | 1,264,006 | 1,423,252 | 1,558,147 | 1,840,611 | 1,971,771 | 2,190,514 |
| Tax | | 242,536 | 266,378 | 364,761 | 376,680 | 429,762 | 483,906 | 529,770 | 625,808 | 670,402 | 744,775 |
| Profit after Tax | | 470,804 | 517,087 | 708,065 | 731,202 | 834,244 | 939,346 | 1,028,377 | 1,214,803 | 1,301,369 | 1,445,739 |
| Net Cash Flow | (13,200,000) | 1,790,804 | 1,837,087 | 2,028,065 | 2,051,202 | 2,154,244 | 2,259,346 | 2,348,377 | 2,534,803 | 2,621,369 | 2,765,739 |

| | |
|---|---|
| *Net Present Value* | $49,551.40 |
| *Profitability Index* | 0.38% |

# Distribution of NPV

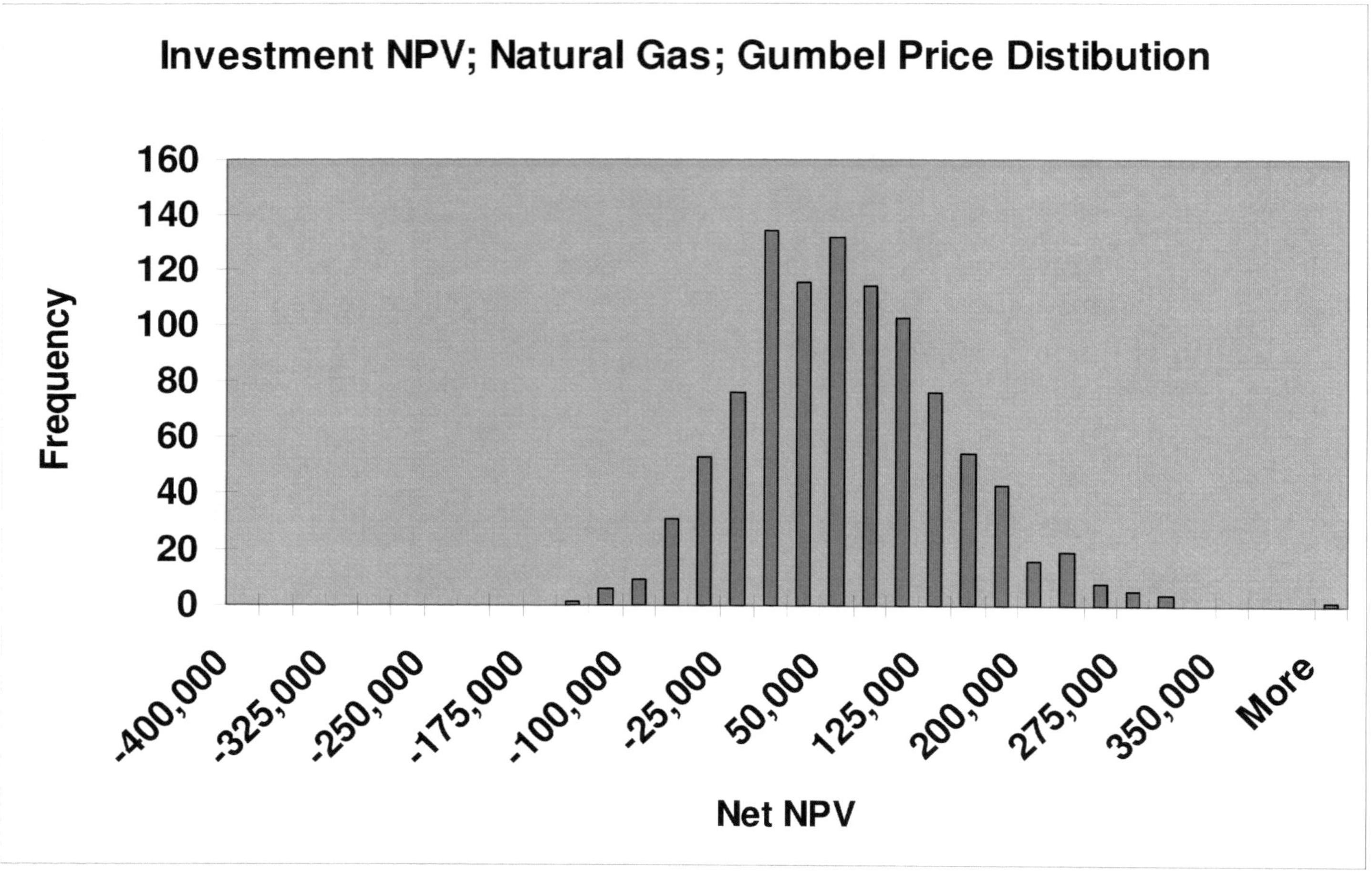


***Expected NPV = $44900; 17% Chance of negative NPV; 15% Chance NPV greater than $100,000***

# Multiple Fuel Option

# Fuel Oil Trend Line

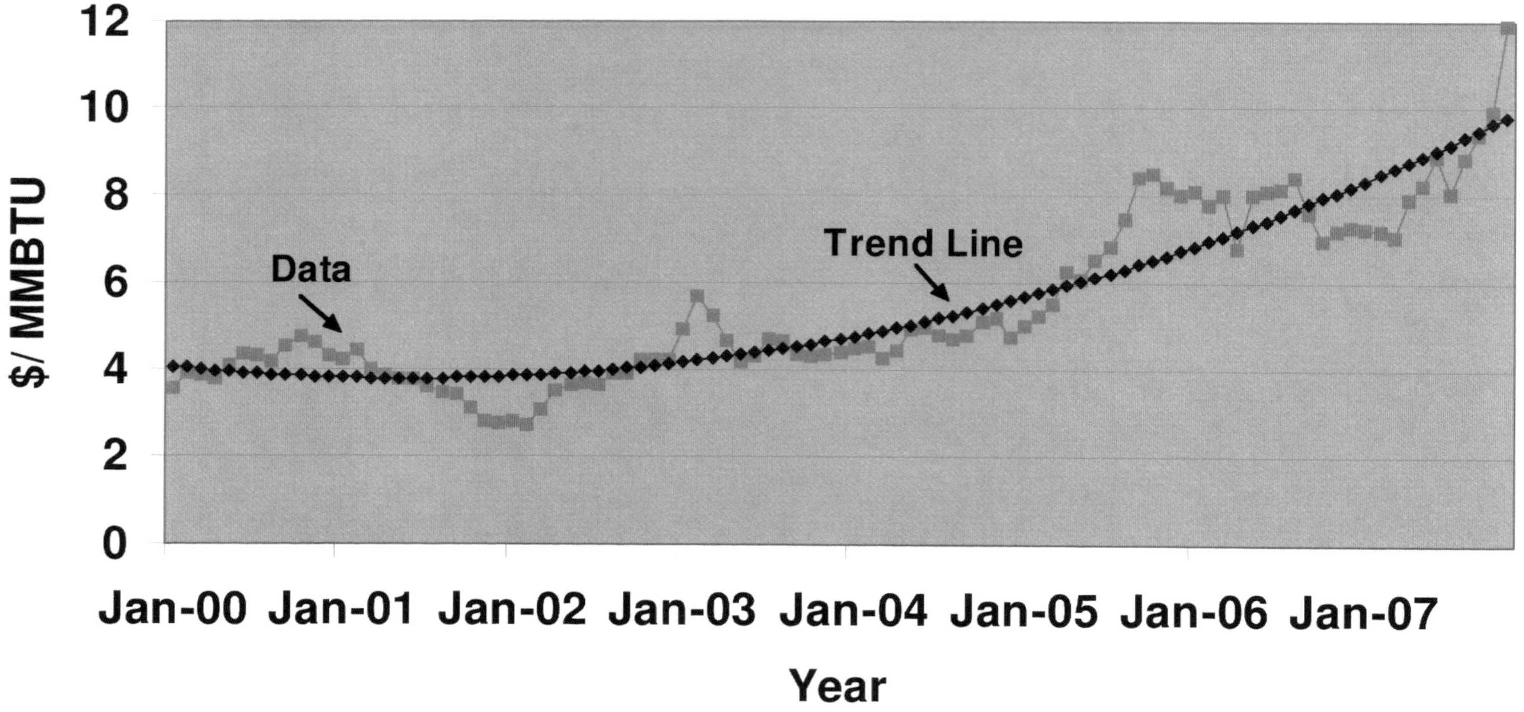

# Fuel Oil – Distribution of Residuals

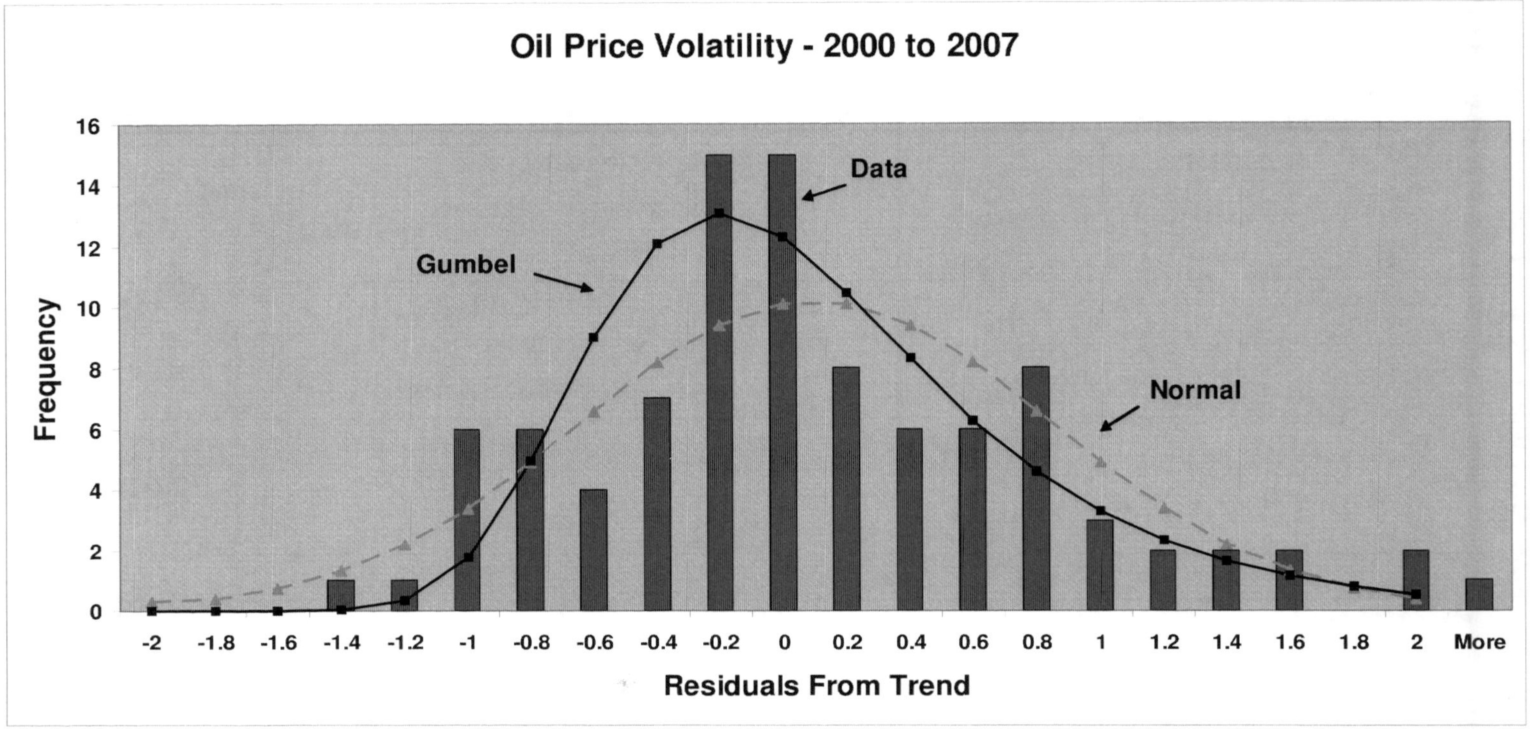

# Assumptions For Trials

**Table 1**

| Assumptions | | | | | |
|---|---|---|---|---|---|
| **Process Demand MMBTU/ Hr** | 150.0 | | | | |
| **New Heater Investment Cost** | $ 13,500,000 | **Maintenance Costs Per Year, % Investment** | 2% | | |
| **Initial Fuel Price, $/ MMBTU** | $ 8 | **Fuel Price Increase Per Year, $/ MMBTU** | 0.56 | **Gumbel Gas Variance** | 0.0435 |
| **Old Equipment, Efficiency, %** | 75 | **Equipment Life, Years** | 10 | **Gumbel Oil Variance** | 0.453 |
| **New Equipment, Efficiency, %** | 90 | **Operating Days/ Year** | 355.0 | | |
| **Fixed Costs/ Yr** | $ 15,000 | **Cost of Capital** | 10% | | |
| **Depreciation/Yr** | $ 1,350,000 | **Tax Rate** | 34% | | |

# Single Trial Result

Table 2

| Year | 0 | 1 | 2 | 3 | 4 | 5 | 6 | 7 | 8 | 9 | 10 |
|---|---|---|---|---|---|---|---|---|---|---|---|
| Average Process Demand MMBTU/ Hr | | 150.0 | 150.0 | 150.0 | 150.0 | 150.0 | 150.0 | 150.0 | 150.0 | 150.0 | 150.0 |
| Fuel Gas Cost, $/ MMBTU | | 7.74 | 8.413406402 | 9.003301645 | 9.6949556 | 10.2731323 | 10.6212811 | 11.2625033 | 11.890622 | 13.119315 | 13.0440308 |
| Fuel Oil Cost, $/ MMBTU | | 8.90 | 8.146812134 | 8.845060446 | 9.18764907 | 10.2070539 | 10.3190481 | 10.5958057 | 12.007154 | 12.216525 | 12.562029 |
| Fuel Cost Savings, $/ Yr | | 2,198,444 | 2,465,120 | 2,601,878 | 2,897,442 | 2,936,336 | 3,102,278 | 3,387,893 | 3,376,937 | 3,982,278 | 3,841,393 |
| Fixed Cost | | 15,000 | 15,000 | 15,000 | 15,000 | 15,000 | 15,000 | 15,000 | 15,000 | 15,000 | 15,000 |
| Maintenance Costs | | 270,000 | 270,000 | 270,000 | 270,000 | 270,000 | 270,000 | 270,000 | 270,000 | 270,000 | 270,000 |
| Depreciation | | 1,350,000 | 1,350,000 | 1,350,000 | 1,350,000 | 1,350,000 | 1,350,000 | 1,350,000 | 1,350,000 | 1,350,000 | 1,350,000 |
| Profit before Tax | | 563,444 | 830,120 | 966,878 | 1,262,442 | 1,301,336 | 1,467,278 | 1,752,893 | 1,741,937 | 2,347,278 | 2,206,393 |
| Tax | | 191,571 | 282,241 | 328,739 | 429,230 | 442,454 | 498,875 | 595,984 | 592,258 | 798,074 | 750,174 |
| Profit after Tax | | 371,873 | 547,879 | 638,140 | 833,212 | 858,882 | 968,403 | 1,156,909 | 1,149,678 | 1,549,203 | 1,456,220 |
| Net Cash Flow | (13,500,000) | 1,721,873 | 1,897,879 | 1,988,140 | 2,183,212 | 2,208,882 | 2,318,403 | 2,506,909 | 2,499,678 | 2,899,203 | 2,806,220 |

| | |
|---|---|
| *Net Present Value* | $62,959.79 |
| *Profitability Index* | 0.47% |

# Effect of Volatility

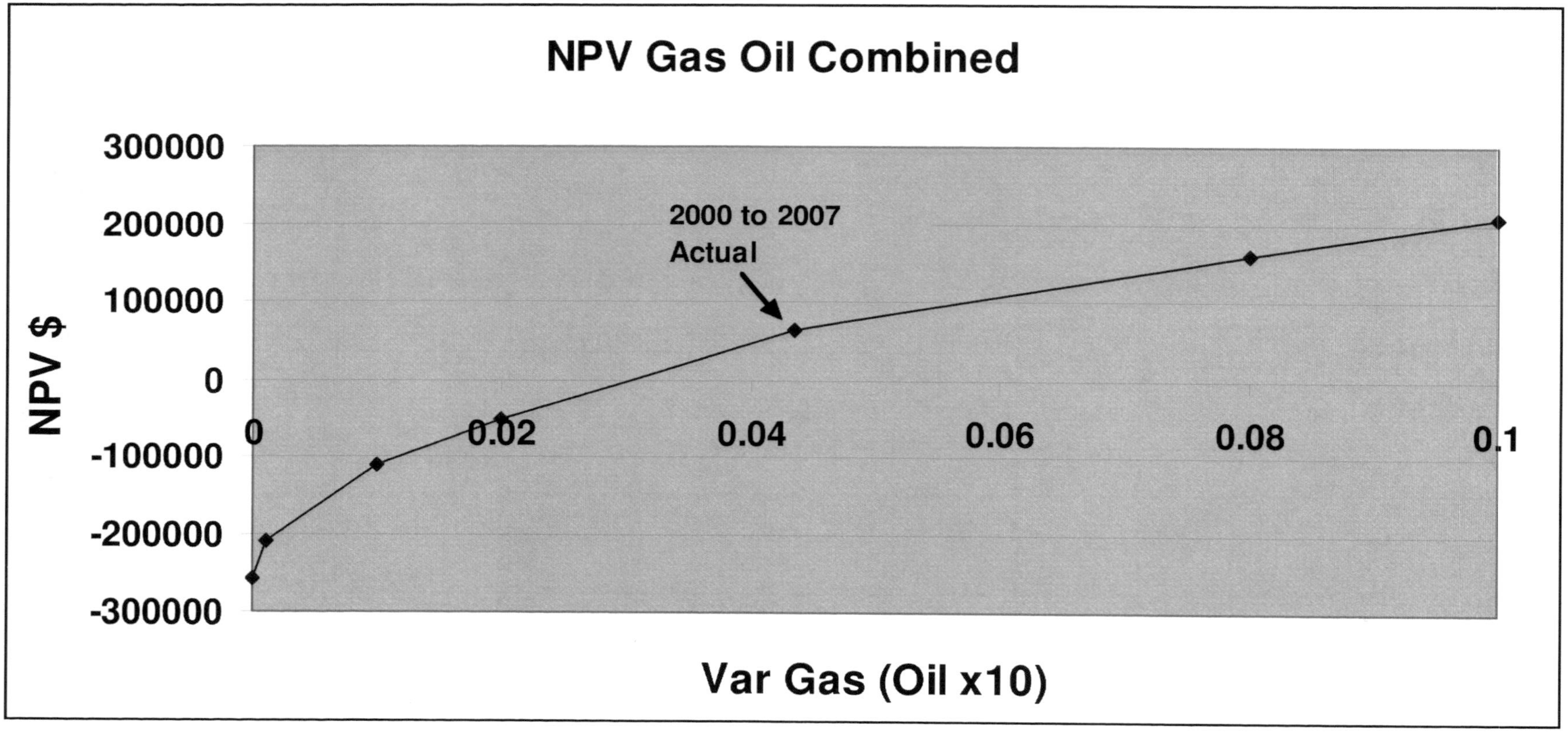


**Expected NPV increases with increased distribution variability!**

# Summary

- Energy price volatility is characteristic of the current and expected future environment
- Monte Carlo simulation is a useful tool for evaluating investment returns with uncertain payouts
- Price volatility significantly affects the expected value of the NPV for alternate fuel investments – more volatility yields a higher expected return

# 164c Managing Greenhouse Gas Emissions Via Co-Utilization of Biomass and Coal: A Compliance Strategy

Pallavi Chitta and **Matthew McArdle**. Mesa Reduction Engineering & Processing, Inc., Auburn, NY 13021

Significant international as well as local regulatory changes are limiting the amount of greenhouse gas emissions. This directly impacts existing and future fossil fuel based energy conversion facilities. The implementation of advanced combustion techniques along with renewable fuel sources in these facilities will be critical in complying with the new restrictive emissions levels. Biomass cofiring is a low cost approach to integrate a renewable energy component in existing coal-fired boilers.

This paper evaluates existing and future projects that co-utilize biomass in advanced injection schemes to assist in CO2 and other emissions reductions. Several issues such as biomass characterization, aggregation and pre-processing schemes would be discussed. Oxygen enriched combustion would be discussed as a possible approach for efficiency gains and process intensification.

# 164d Effect of Availability on Multi-Period Planning of Oil and Gas Production Systems

**Karla L. Ruiz**[1], M. Sam Mannan[1], and Richart Vazquez Roman[2]. (1) Chemical Engineering, Mary Kay O'Connor Process Safety Center, Artie McFerrin Department of Chemical Engineering, Texas A&M University., College Station, TX 77843-3122, (2) Chemical Engineering, Instituto Tecnológico de Celaya, Departamento de Ingeniería Química., Av. Tecnológico s/n, Celaya, Gto., CP 38010, México., Mexico

Natural gas and petroleum are non-renewable and scarce energy sources. Although, it is well known that hydrocarbon reserves are depleting through the years, oil and gas remain the principal source of energy upon which our society is strongly dependent. Hence, the optimization and accurate planning of hydrocarbon production are the main keys to making it safer, more efficient, and cheaper. One of the tools commonly used to evaluate the optimization of oil/gas production system is the process simulation modeling.

A hydrocarbon production system typically consists of at least one underground reservoir where several wells have been drilled into the hydrocarbon-bearing rock to conform a fixed topology network. Wells are interconnected with manifolds to transport the gas or oil to a storage or sale location. The process simulation consists of calculating the total hydrocarbon production for the given production system. The pressure in the wellbore is the main variable to determine in the process. When oil/gas is produced, this pressure decreases until the production cannot be sustained. Then, the production decreases until the wells stops producing. If the well is shut down, the pressure at the wellbore increases because of the natural gas flow coming from the reservoir. In addition, artificial lift techniques, such as water injection, gas lift and pump systems can be incorporated into the simulation program. The oil/gas production has been also modeled as a multi-period optimization case to incorporate the possibility of different demands, cost and overall time behavior. The current field optimization approaches do not take into account the availability of the equipment, adding to the planning a lot of uncertainty.

In this work, we have integrated the availability analysis to the above model description. The availability of a system is analyzed by Monte Carlo simulation, which involves the modeling of the probabilities of failure, the type of failure, the time to repair associated with each failure, and time of occurrence for a field system. As a result, a new production planning is accomplished in the effective work period. It is known that availability is the ability of an item to perform its required function over a stated period of time. Thus, an accurate measure of the availability will reduce the uncertainty on a multi-period planning production of either oil or gas. Furthermore, this efficient approach will have beneficial use in financial risk decisions.

# KBC Helps Chevron Phillips Cut Energy Costs

## Chevron Phillips Chemical Company LP

*Port Arthur, Texas*

Prepared by: Chuck Polito, Chevron Phillips Chemical Company LP
Chris Bealing, KBC Energy Services
Allan Rudman, KBC Energy Services

AIChE Ethylene Producers Conference 2008

*January 2008*

KBC ADVANCED TECHNOLOGIES, INC.

***Proprietary Information***

By accepting this document, the recipient confirms that all information contained herein will be kept confidential, and this information will not be disclosed to third parties without the prior written agreement of KBC Advanced Technologies, Inc and Chevron Phillips Chemical Company LP.

***KBC Advanced Technologies, Inc.***
14701 St Mary's Lane
Suite 300
Houston, Texas 77079
USA

T +1 281 293 8200
F +1 281 293 8290

# Chevron Phillips Cuts Energy Use with a KBC Strategic Review

## 1.0 Abstract

This paper describes the application of a KBC Strategic Energy Review (SER) at the Chevron Phillips Chemical Company LP (CPChem) site in Port Arthur, Texas. The full SER program combines benchmarking, operational reviews and Pinch analysis with a systematic Total Site approach to define a complete spectrum of energy benefits ranging from in-process improvements to the optimization of the site-wide utility system.

New opportunities arising from the SER were combined with projects already under consideration by CPChem into a single energy improvement RoadMap which considered the strategic decisions for energy improvement and the interactions between the proposed projects to define a coherent forward implementation and investment plan. The analysis of these interactions was performed using a validated site utility model that was jointly agreed upon by KBC and CPChem.

Projects recommended for rapid implementation in the upcoming turnaround significantly reduced energy consumption while achieving an overall payback period of one year. Medium-term investment opportunities then doubled the total energy savings with an incremental payback of less than two years.

## 2.0 Introduction

Chevron Phillips Chemical Company LP (CPChem) manufactures a range of petrochemical products including olefins, polyolefins, aromatics and styrenics at its major plants located throughout the Gulf Coast, including Sweeny, Cedar Bayou, Port Arthur, and St. James. These are world-scale operations with significant demands for heating and power. CPChem believes there is an opportunity to improve the energy performance of these sites and was therefore interested in doing an energy benchmarking audit. The Port Arthur site was chosen first because of the need to solve the problem of the Cyclohexane unit venting low pressure steam and also to validate the energy improvement items identified by the plant process audit team for the upcoming 2008 shutdown.

In July 2007, the KBC Energy Services (KBC) team completed a ten-month two-phase study Strategic Energy Review (SER) at Port Arthur. This SER combined experience-based methods with a systematic Total Site approach to define a complete spectrum of energy benefits ranging from immediate return to longer-term investment projects.

The following diagram summarizes the full SER and shows how the work was split between Phase 1 and Phase 2:

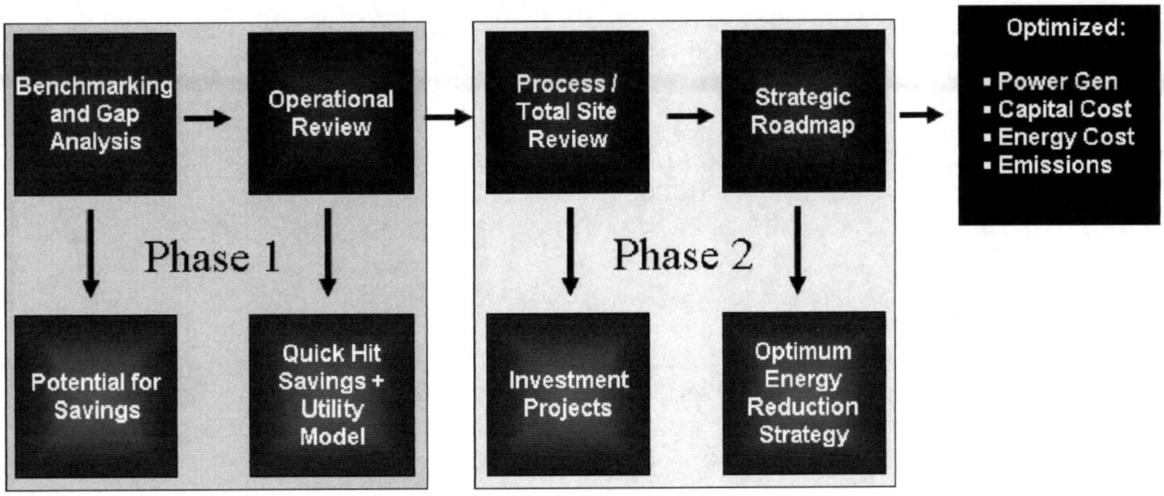


As part of the Phase 1 work, the Port Arthur site was benchmarked against other similar sites worldwide using the KBC Best Technology (BT) index and gaps were identified between actual and optimized site energy performance.

The major focus of Phase 2 was then to define projects to close these energy gaps leading to the production of a strategic RoadMap to provide a coherent forward implementation and investment plan for the Port Arthur site. Also of interest was the determination of a clean versus dirty operational gap to be addressed during the upcoming turnaround.

## 3.0 Site Overview

The Port Arthur site comprises a world-scale Ethylene plant along with Cyclohexane and Cumene Feed Preparation units and a third-party Cumene plant. The site imports steam at two pressure levels from a neighboring plant to supplement its on-site generation.

## 4.0 Best Technology Benchmarking

The KBC BT index is defined as the ratio of the total energy consumption (including fuels, power and steam import) to the sum of BT energy consumption allowances (standards) for each process unit and associated infrastructure.

KBC BT standards have been developed by fundamental analysis and design studies on individual process units. They take into consideration the key process parameters affecting energy consumption along with the specific site economics and a first quartile BT index is achievable in a grass roots unit built today.

The calculated BT for the base case operation at Port Arthur site including both Ethylene and Cyclohexane units placed the site on the border of the third and fourth quartiles in KBC's extensive petrochemical database.

## 5.0 Utility System Model

A model of the site fuel, steam and power system was constructed in KBC ProSteam software including all operating constraints. The model was used to simulate the impact of project changes in terms of the marginal utilities at site boundary. This ensured that the true cost savings were calculated for each project and that interactions between projects were fully addressed. At Port Arthur, the main marginal utilities were exported fuel gas and imported electrical power and steam purchased from third-parties.

The model included simulations of all the individual process units and the links between them for four different operating modes. These operating modes reflected changes in feed slate for the Ethylene plant and variations in capacity on Cyclohexane. The overall saving was calculated as a weighted average across these four modes.

The base model simulated a theoretical clean operation for project evaluation that was validated and agreed upon by KBC and CPChem. It therefore represented a case where several planned easy-win opportunities had already been implemented. Several of these opportunities related to the correction of problems related to the maintenance of plant equipment and some related directly to the impact of Hurricane Rita in 2005, which caused an unplanned shutdown for an extended period of time due to cooling tower damage. Using a base case which assumed that the plant was running well meant that CPChem could be confident that the proposed RoadMap would be applicable to the future operating condition of the site. It also increased the challenge to find improvement projects by taking much of the low-hanging fruit out of the scope of the project.

The following diagram shows a simplified diagram of the Ethylene plant steam system based on the model constructed in ProSteam.

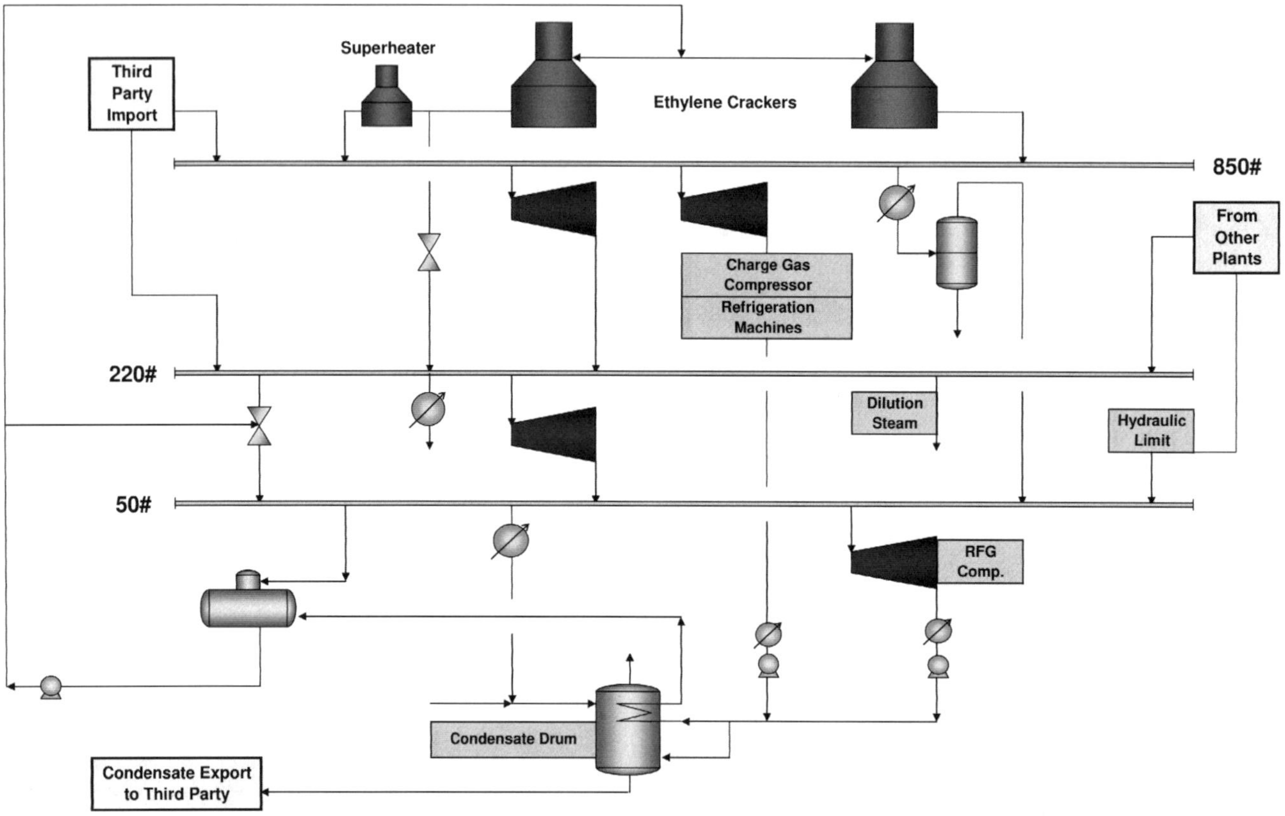

## 6.0 Steam Balance Issues

A major inefficiency on the Port Arthur site was the venting of 50# steam. Cyclohexane is an exothermic process and is therefore a net generator of steam. With the third-party Cumene plant also exporting at the 50# level the site typically saw a significant excess.

In most operating modes, this problem was increased by a hydraulic limitation on the single undersized transfer line which can take 50# steam from Cyclohexane to the Ethylene plant. Because of this, the site found itself in a situation where steam was being let down from the 220# level into the 50# header to make up the shortfall on Ethylene, while simultaneously venting a significant amount of 50# steam on Cyclohexane. Correcting this situation would become one of the main issues in the final RoadMap.

## 7.0 Quick-Win Phase 1 Projects

The phased approach of the SER is designed to maximize the efficiency of the KBC project team. Phase 1 looks across the full site and identifies potential areas for improvement, while Phase 2 then hones in on these areas to develop the detailed project descriptions and implementation RoadMap.

In carrying out Phase 1, any opportunities for quick-win projects are identified and reviewed. This enables the site to realize immediate benefits – even before the project is completed. At Port Arthur, it was possible to convert a steam-driven spare boiler feed water pump from slow rolling to hot standby by utilizing better steam trapping and new electronic governors, while still maintaining the reliability of the critical pumping systems. A further opportunity was to bypass a product cooler, which allowed the downstream plant to be fed hotter.

Another key part of Phase 1 is a mechanical walkthrough where a KBC specialist reviews the process units and assesses the operation of the actual equipment. This review focuses on areas such as pump and turbine operation, cooling water systems, major compressors and heater performance. The walkthrough identified several opportunities for improvement, many of which could be implemented immediately. The main findings included the use of temperature scans on exhaust steam lines to identify inefficiencies in the extraction turbines used for boiler feedwater, quench water and process pumping. These inefficient turbines which utilized excessive amounts of medium pressure steam have been identified for maintenance overhaul during the upcoming turnaround and are expected to result in significant energy savings once they are brought back to design conditions. Additional findings included more efficient operation of the cooling tower with better water distribution and optimum cooling tower fan operation. The mechanical walkthrough also provided some insight on the plant power factor, its impact on higher utility costs and capacity limits on motor drivers, particularly the cooling tower fans and furnace ID fans.

## 8.0 Energy Improvement Roadmap

In developing the future energy strategy for the Port Arthur site, the overriding factors which determined the order in which projects were proposed for implementation were the project economics and their relative ease of installation and operability. In categorizing the projects based on these factors however it was necessary to consider the impact on the overall site steam balance to minimize both steam venting and letdowns. The projects developed during Phase 2 of the SER were grouped as follows:

| | |
|---|---|
| Group 1 | Projects planned for implementation in the forthcoming shutdown |

| | |
|---|---|
| Group 2 | Other low cost or quick return projects recommended for rapid implementation |
| Group 3 | Medium/longer payback projects or projects requiring further investigation. Not proposed for rapid implementation |
| Group 4 | Long-term strategic investment |

The following sections describe in broad terms the projects included at each stage in the RoadMap.

## 8.1 Group 1 Projects

Following Phase 1 of the SER, CPChem and KBC developed a list of opportunities which were planned for implementation in the forthcoming shutdown. This list comprised a combination of projects that came out of the Phase 1 work and projects that were already under consideration by CPChem.

A selection of the Group 1 projects, along with a brief description of each opportunity, is listed below:

| **Plant** | **Description** |
|---|---|
| Ethylene | Install secondary TLEs on the four newer cracking furnaces |
| CFPU | Improve Deethanizer analyzer control to reduce reboiler steam |
| Ethylene | Performance improvements for propylene compressor, minimizing kick-back by utilizing full exchanger areas |
| Ethylene | Improved damper control on 4 cracking furnaces and common economizer for the older 12 cracking furnaces. |
| Ethylene | Stop slow rolling spare quench water pump |

As discussed earlier, the impact on the 50# steam balance has been seen to be a critical issue for Port Arthur both in terms of venting and in terms of the amount of 220# steam being let down to 50# on the Ethylene plant. Implementing these Group 1 projects has the following impact:

| **Weighted Average Flows** | **Base Case** | **After Group 1** |
|---|---|---|
| 220-50# letdown | 9 klb/h | 10 klb/h |
| 50# vent | 44 klb/h | 44 klb/h |

This table shows that the projects actually serve to increase the letdown slightly but the increased 50# demand caused by the secondary TLE project is largely balanced out by the other proposed changes. The benefits from the TLE project is not shown on the table, of which the project resulted in a significant amount of reduced imported 850# steam.

The table also shows that the Group 1 projects have no effect on the 50# vent. The reason that the vent is unaffected is because it occurs on the Cyclohexane plant and no improvements have yet been proposed in that area.

## 9.0 Site-Wide Integration

The venting problem on the Cyclohexane plant cannot be solved by looking at the plant on a stand-alone basis. In fact, the opposite is true; KBC work identified heat exchanger network modifications both to reduce steam consumption on the plant and increase steam generation and these would increase venting unless a solution could be found.

It is a common finding from KBC work on similar projects throughout the world that a lot of effort has been put into improving the energy performance of process plants but that the focus of this work has been almost exclusively within the individual plant boundaries. A strength of the KBC Total Site approach is that plants are considered in a site-wide context alongside the supporting utility system. This allows the best overall strategy for the site to be defined both in energy saving and in capital investment terms.

Returning to the venting issue at Port Arthur, the requirement from a site-wide perspective was to find a location for the excess steam to be used. The difficulty with this was that the steam was available at the 50# level. The fact that a letdown remained between the 220# and 50# levels on the Ethylene plant meant that there was scope to use part of the excess steam provided that the export line from Cyclohexane was upgraded. However, as the results after the Group 1 projects showed, the letdown flow was relatively small at 10 klb/h so exporting this amount of additional steam from Cyclohexane would only reduce the vent rate from 44 klb/h to 34 klb/h as shown in the following diagram. Although there was still a pay-out to phase in a new line to eliminate the letdown with the "zero" valued steam that would otherwise be vented, a more fundamental solution to the problem was required.

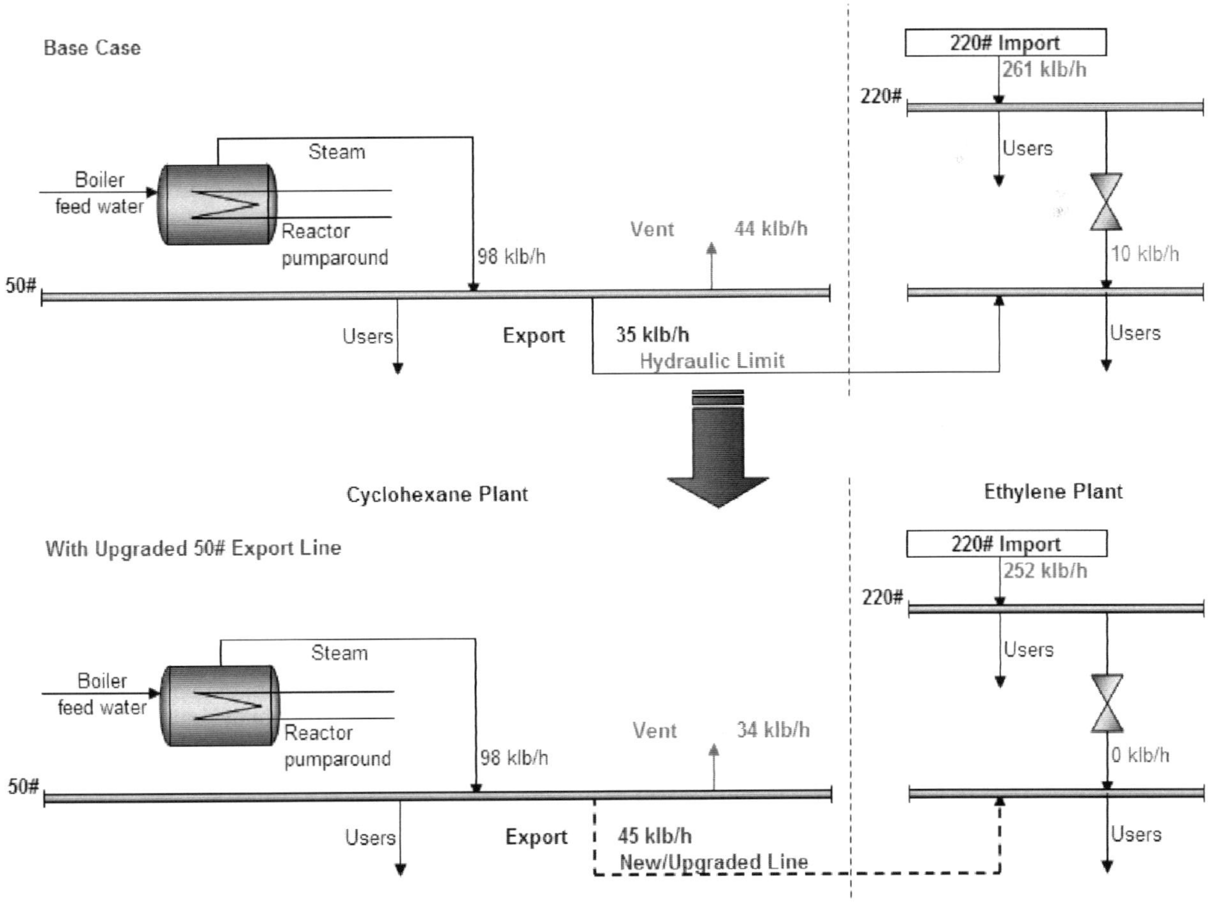

Looking again at the Cyclohexane plant showed that the steam was actually generated at around 100# pressure before being letdown internally into the 50# header. The possibility was then considered of installing a new steam line to enable this intermediate pressure steam to be exported from the unit. The problem then became finding a suitable use for the steam.

The only possible sink that had a reasonably high demand was the Ethylene plant dilution steam system. In the base case, dilution steam was supplied directly from the 220# header but the required steam pressure was significantly lower. Allowing for pressure drop, it was found that it would be possible to supply steam to twelve of the cracking furnaces on the Ethylene plant that typically run at lower coil operating pressures. Because of the fluctuations in plant operation, the system was set up to cope with situations where either there is an excess of Cyclohexane steam which needs to be let down or there is a shortage of Cyclohexane steam which needs to be made up from the 220# header. This included making provisions for times when Cyclohexane is shutdown for turnaround and/or maintenance.

One further issue was the reduced heat content of the dilution steam in moving from superheated 220# steam to saturated 110# steam or a mixture of the two. To avoid possible problems associated with poor dilution steam flow measurement due to two-phase flow, a new heat exchanger was proposed using 220# steam to heat the dilution steam. The impact on the steam balance of this project is shown in the following diagram:

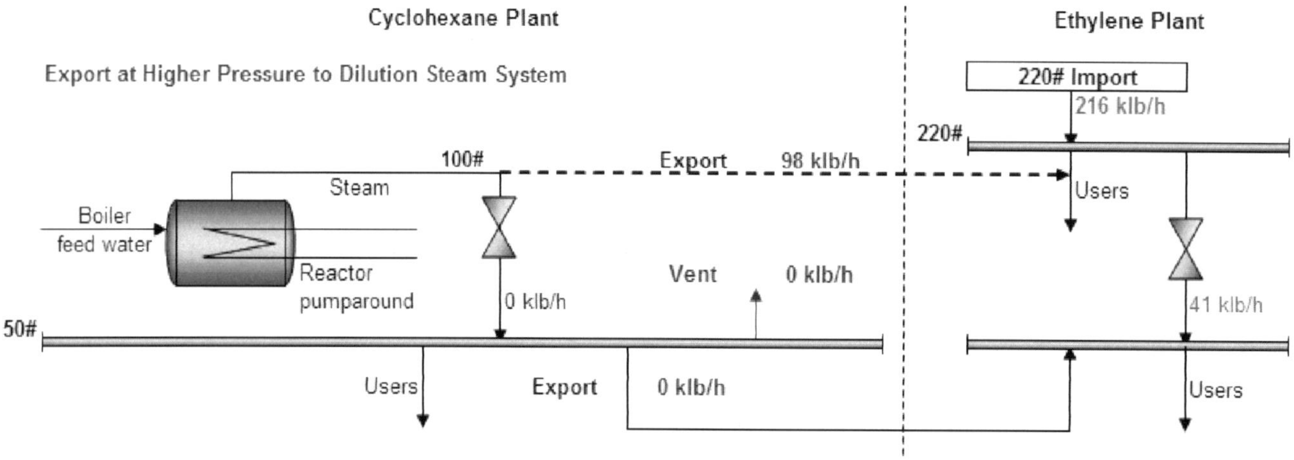


A major impact of this project is that in eliminating 50# venting it moved the site into a position where there was a significant deficit of 50# steam which needed to be made up by the 220-50# letdown. The above diagram shows how the 220-50# letdown flow had increased significantly because steam was no longer imported at the 50# level and this negated some of the benefit from the new steam export line. However, when the project was considered in the context of the RoadMap, it now meant that 50# steam saving projects became attractive as they would reduce the letdown flow giving a direct saving in 220# import. The overall goal was to reduce this import by a flow equivalent to the amount of Cyclohexane vent steam used as dilution steam. This again emphasizes the importance of a credible overall site steam model to highlight the interaction of projects being phased into the RoadMap.

## 10.0 Energy Improvement Roadmap Continued...

After identifying a solution to the main inefficiency in the steam system, the development of the RoadMap was taken further.

## 10.1 Group 2 Projects

The Group 2 projects represent those that were recommended for rapid implementation but were not included in the shutdown list because of time constraints at the end of the study.

The Group 2 projects, along with a brief description of each opportunity, are listed below:

| Plant | Description |
|---|---|
| Ethylene/Cyclohexane | Export dilution steam from Cyclohexane |
| Cyclohexane/Cumene | Import 50# steam and condensate into Cyclohexane |
| Ethylene | New condensate vs. deaerator feedwater heater |

The site-wide interactions involved in the export of dilution steam from Cyclohexane are discussed above but this project also opened up opportunities to improve integration between Cyclohexane and the third-party Cumene plant. By importing 50# steam into Cyclohexane, it was possible to supply the Cyclohexane 50# consumers without needing to let steam down from its 100# generation level. Also, by bringing hot condensate in as feed to the Cyclohexane deaerators without an economic penalty, it was possible to reduce the demand for deaeration steam. Both of these opportunities allowed the amount of 100# steam available for export to the Ethylene plant to be increased.

Also, in order to minimize venting of the low pressure condensate flash drum on the Ethylene plant, a new condensate/deaerator feedwater heater was proposed. This would in turn minimize 50# steam requirements on the deaerator.

Implementing these projects has the following impact on the 50# system:

| Weighted Average Flows | After Group 1 | After Groups 1+2 |
|---|---|---|
| 220-50# letdown | 10 klb/h | 28 klb/h |
| 50# vent | 44 klb/h | Zero |

## 10.2 Group 3 Projects

The Group 3 projects are those that are proposed for medium-term implementation. There are various reasons why projects have been included in this category as opposed to Groups 1 or 2. Many of the projects require more detailed engineering work in order to finalize the designs and therefore would not be ready for rapid implementation. A summary of the Group 3 opportunities is given in the following table:

| Area | Description |
|---|---|
| Refrigeration Pinch Projects | Three network revamp projects in the Ethylene plant |
| Turbines/Motors | Replacement of two turbines with motors in the Ethylene plant<br>Replace condensing turbines with back-pressure in cooling water system |
| Heat Exchangers | Three revamp opportunities on hot end of Ethylene plant |

Of particular interest at this stage is the choice of turbine and motor options. It is critical at this point that the selection made fits with the overall steam balance and also retains flexibility to react to process fluctuations. The turbine/motor and heat exchanger projects help to better balance the 50# system by further reducing the 50# demand while simultaneously increasing its supply thereby minimizing 220-50# letdown.

There were three high pay-out refrigeration pinch projects were identified that helped reduced the HP steam demand. CPChem identified one of these projects during the plant process audit. KBC validated its payout and also identified two other high payout refrigeration pinch projects using Pinch analysis in the Ethylene plant.

As at other stages in the RoadMap, we can look at the impact on the letdown and venting of steam:

| **Weighted Average Flows** | **After Groups 1+2** | **After Groups 1+2+3** |
|---|---|---|
| 220-50# letdown | 28 klb/h | 3 klb/h |
| 50# vent | Zero | 7 klb/h |

The above table shows that there is a small average amount of letdown and vent. It is worth noting that this is the average across four different operating modes and that venting and letting down does not occur simultaneously. The results also show that the system is now well balanced in achieving such low averages across widely varying conditions.

### 10.3 Group 4 Project

Group 4 in the RoadMap is for projects which are seen as long-term strategic measures for the Port Arthur site and that do not currently offer attractive paybacks. Following discussions with CPChem, only one project was deemed to fall into this category and this related to extensive modifications to four of the crackers. CPChem has elected to put this project on hold indefinitely due to operational issues and its high complexity. Future environmental regulations may drive CPChem to implement these projects in the future.

## 11.0 Impact On Site BT Index

The proposed projects move the Port Arthur site from a position in the on the border of the third and fourth quartiles to one at the top end of the second quartile.

The question of why it has not been possible to move the site into the first quartile then needs to be asked. This comes down to understanding what further improvement measures would be necessary to the site energy system. Possible further improvements that would reduce the BT at Port Arthur include:

- VHP steam generation in TLEs
- Gas turbines including possible direct integration into crackers
- Large turbine modifications
- Installation of HP or VHP on-site boiler system with air preheaters

The problem with all these ideas are that they are highly capital intensive and difficult to realize in a retrofit situation. On the whole, the new plants need to be capital intensive, requiring high energy costs for justification.

## 12.0 Conclusions

A total of 22 projects were included in the full RoadMap. These projects were split into stages as follows:

| | |
|---|---|
| Group 1 | Projects planned for implementation in the forthcoming shutdown |
| Group 2 | Other low cost or quick return projects recommended for rapid implementation |
| Group 3 | Medium/longer payback projects or projects requiring further investigation. Not proposed for rapid implementation |
| Group 4 | Long-term strategic investment |

The Group 1 and 2 projects combined achieved savings of several million dollars with an overall payback of just one year. The Group 3 projects then doubled the savings potential with an incremental payback of less then two years.

The Group 3 projects are seen as medium-term opportunities while Group 4 encompasses the major revamp of some of the crackers which is seen as a long-term option.

## 13.0 About the Authors

**Chuck Polito** is a Senior Olefins Technical Engineer for Chevron Phillips Chemical located at Cedar Bayou, TX. He has executed reliability, process improvement, and energy improvement projects for both the Port Arthur and Cedar Bayou olefin units. His focus areas have recently been on capacity and energy improvement on cracking furnaces, distillation and refrigeration areas inside the ethylene plant.

**Chris Bealing** is a Senior Staff Consultant within the KBC Energy Services group. He joined the company in 1993 and has extensive experience in the application of all aspects of energy improvement technology having carried out numerous projects across a range of process industries. He can be reached by email at cbealing@kbcat.com.

**Allan Rudman** is the Vice President of the KBC Energy Services group, resulting from the acquisition of Linnhoff March in 2002. He has executed numerous performance and profit improvement projects in process industries, particularly oil refining, petrochemicals, general chemicals and specialty chemicals. Utilising his engineering background, he has played a leading role in the development of KBC Energy Services' proprietary tools, including training, planning and methodology guidelines.

## 201a The Louisiana Petrochemical Megasite Project

**Dane Revette**, Louisiana department of economic development, New Orleans, LA 70130

The Louisiana Petrochemical Megasite Project is an initiative funded by Louisiana Economic Development (LED). This project is being undertaken by the Center for Business and Information Technologies (CBIT) at the University of Louisiana. This project is designed to lay the Foundations for a State initiative establishing a Louisiana Petrochemical Megasite. The Project has thus far, examined nine Megasites around the world. It has also surveyed the petrochemical industry on its joint venture experiences. From these studies, together with our industry partners, the factors fostering Megasite success have been distilled. The project then analyzed the infrastructure and policies of the State to see how Louisiana compares. Currently, the team has an economic feasibility study underway and is preparing a policy and implementation roadmap for establishing a Petrochemical Megasite in Louisiana.

## 201b Smart Plants and Smart Management Prevent Plant Problems

**Jimmy L. Humphrey**, J. L. Humphrey & Associates, 3605 Needles Dr., Austin, TX 78746, J. Christopher Lewis, Technology R& D, ConocoPhillips, 166 PDC Bartlesville Technology Center, Highway 60 & 183, Bartlesville, OK 74004, and A. Frank Seibert, Separations Research Program, University of Texas at Austin, 10100 Burnet Road, Bldg 133/CEER, Austin, TX 78758.

Smart Manufacturing Plants and Smart Management prevent plant problems before they occur. Examples of problems which can be prevented include unplanned shutdowns, excessive energy use, unwanted safety events, and environmental emissions. Accurate real time plant measurements and data are the foundation of the Smart Plant. In addition to establishing current status of the plant, these data are used in models to PREDICT future plant performance. Selected results from the predict step are used to PRO-ACT where proactive actions are taken to optimize future plant performance. The desired end result of the pro-act step is to PREVENT plant problems before they occur.

# MANAGING THE ENVIRONMENT WITH "SMART PLANT" INNOVATIONS

***Author: Jeremiah O'Brien, Process Data Control Corp, Arlington, Texas***

"Smart Plants" are, by definition, sustainable and efficient due to the implementation of effective instrumentation and controls, work processes, and management systems. Specific innovations may include instrumentation and control systems that provide new insights into current and future operational conditions, as well as more effective field instruments, networks, process historians, and process simulation software. While these types of innovations cannot make plants any smarter, they can make the staff more aware, capable, and efficient and, consequently, more successful.

But, how can environmental management be enhanced by the implementation of "smart plant" technologies?

## Objectives of "Smart Plant" Technologies

To understand the relationship between environmental management and "smart plant" technologies, it is important to examine the reasons why companies implement these technologies in the first place.

### *Efficiency*

Many "smart plant" technologies are aimed squarely at improving efficiency and, hence, reducing the cost of production. To achieve this objective, the innovations must result in the same, or improved, output with reduced investment of time, effort, or other inputs. For example, a Gulf Coast area refinery implemented a data acquisition system to track "routine duties" of operations staff that replaced log sheets, checklists, and other paper-based methods with hand-held computers. Although a range of goals and objectives were involved in this decision, the major "driver" was efficiency – not just in data collection but also in the processing and downstream utilization of data.

### *Safety*

"Smart plant" technologies that involve simulations and forecasting are often implemented to reduce the risk of malfunctions and unexpected outages of production equipment. Better-informed operating personnel tend to make better decisions. Improvements in the decision-making of operators, unit supervisors, production managers and planners, and other staff can enhance safety in a variety of ways.

### *Reliability*

"Smart plant" technologies may be deployed to increase the reliability of plant processes and the quality of finished products. This objective relies on the well-accepted premise that quality and reliability can be improved through greater conformity with established standards. Non-conformance with standards is often associated with low quality, so any technology or approach that enhances conformity has the potential to increase the quality of an organization and its products.

"Smart plant" technologies strive to use data – possibly data that was not previously collected – to inform and educate company staff. Standard procedures and "best management practices" enable the better-informed staff to act and react to conditions more effectively. The result is more consistency in all phases of production which, in turn, leads to more reliable outcomes.

***Knowledge Management***

"Smart plant" technologies often utilize computer systems to capture and reuse knowledge that would otherwise be known to only a few members of the staff. With record numbers of engineers and operators in today's process industries being awarded early retirement, it has never been more important to capture the knowledge of skilled and experienced workers and apply this knowledge to solving new problems.

The need for more knowledgeable workers is intensified by the global economy, in which new plants are often built in areas where the local work force has little or no experience or skills in critical areas, such as operations. Successfully deployed "smart plant" technologies can accomplish this goal by collecting and storing data, extracting knowledge from these data sources, and enabling the staff to access and apply information assets where needed to improve plant operations.

## "Smart Plant" Objectives and Environmental Management

The objectives of "smart plant" approaches are aligned with many of the goals of an Environmental Management System (EMS), as discussed below:

***Efficiency***

Environmental departments share the need to "do more with less" with all other departments in process plants. Efficiency is a major factor in determining not only how much work can be performed with available resources, but also how accurate and complete the work products are when they are produced. Compliance requirements involve both recurring tasks, such as *quarterly* HON Periodic Reports, and *annual* Emission Inventories. However, most applicable rules must be complied with, either continuously or during specified operating conditions, such as during a unit start-up. Thus, efficiency extends beyond determining compliance and maintaining records with minimal effort; it also includes efficient acquisition of data needed to assess what compliance requirements were in effect during specific time periods. Many, if not most, regulatory violations in process plants can be directly attributed to inefficiencies in the timely collection and/or interpretation of information.

***Safety***

Safe plants have a high likelihood of being compliant with applicable environmental rules and regulations. Safety does not necessarily guarantee compliance, but plants that are not concerned about the safety of their equipment or processes probably apply a similar lack of concern to meeting their regulatory requirements.

Safe plants are often the product of attention to detail in the planning and performance of work processes – even to the point of defining written standard operating procedures. Organizations that place importance on standard procedures and best management practices are much more likely to embrace a similar approach to defining standard compliance procedures for reporting, recordkeeping, monitoring, and testing.

Compliance programs must define all of the tasks that are required to meet regulatory requirements, and assign them to individuals who are capable of carrying out their responsibilities in an effective manner. Plants that take safety seriously have developed a similar approach to ensuring that procedures have been defined to reduce unsafe results, and staff has been sufficiently informed and trained – operating principles that readily apply to environmental compliance planning.

***Reliability***

Compliance requirements must be met under virtually all conditions up to and including plant emergencies. In fact, some requirements are applicable only when emergencies occur (e.g., excess emissions reporting). Consequently, reliable systems for acquiring, interpreting, and taking action on compliance-related information are not simply preferable; they are necessary. "Smart plant" technologies, especially those tap into on-line systems and databases, such as process historians, are among the most reliable approaches to meeting these objectives.

***Knowledge Management***

Some compliance requirements demand knowledge of plant design, operating characteristics, maintenance history, and other information that cannot be obtained through "real-time" data collection. For example, regulations dealing with fugitive emissions and continuous emission monitors require quality control programs with long-term data and analysis of that data. "Smart plant" technologies include computer-based approaches to capturing and reusing knowledge that might otherwise be known to only to a few members of the staff. Thus, a "smart plant" is usually a plant with an excellent "memory" for facts that are important to successful operations. To the extent that such facts include information that is needed to assure compliance, then a "smart plant" is more likely to be a compliant plant, too.

## A Working Definition of "Smart" Environmental Management Systems

How can the movement toward "smarter" plants enhance the development of more compliant plants at the same time? The foregoing discussion would tend to confirm that, from the standpoint of meeting objectives, "smart" plants should contribute to the development of a more effective EMS. In this section, we investigate how these concepts can be implemented in a practical approach to environmental management.

### ***Features of "Smart" Environmental Management Systems***

The basic structure of a "smart" compliance program includes all of the following:

- Detailed, equipment-specific lists of applicable rules and tasks.
- The use of "generic task" definitions to populate equipment-specific task lists, both to standardize on preferred or "approved" approaches and to increase efficiency by "broadcasting" the same generic task definitions to all equipment subject to the same rules.
- An automated system for producing forms and data compilations of all kinds as required by regulatory agencies in order to ensure efficiency and responsiveness with such requirements.
- An automated system to record compliance task assignments, with identified "backup" persons, for those tasks that will be performed by people.
- Integration of on-line data repositories, such as process historians, with compliance systems to maximize "self-certification" of compliance tasks using available data, e.g., flow rates, temperatures, etc. This integration should include both evaluating compliance status in light of regulatory limits, if any, and documentation of compliance data, and should include automatic reply of queries that failed to produce definitive data for evaluating compliance.
- Effective tools and techniques that are applied to grouping compliance tasks for effective recordkeeping, including the capability to handle new tasks and modifications to existing tasks in the future.
- An automated system to send task reminders to staff members when their tasks are pending or overdue.
- An automated system to inform staff members and managers when they have entered information task completion data that indicates potential compliance deviations so they can correct or enhance the information as needed and, by doing so, facilitate more timely, accurate, and complete deviation reporting to regulatory agencies.
- The capability to query compliance data efficiently and reliably, to determine which task completion results must be reported to agencies, and/or more fully documented in order to satisfy pertinent regulatory requirements.
- A continuous improvement program that applies the "lessons learned" from operating the EMS, in order to identify and implement improvement to the system.

### ***Essential Processes of "Smart" Environmental Management Systems***

The development of a "Smart" EMS depends on the capability of the EMS to address certain work processes that are common to all EMS in the process industry. These work processes are discussed below:

- Planning – the operation integrity of process plan operations could not be assured without detailed simulation and planning. Likewise, an EMS relies on analysis and planning to be more efficient in the development and maintenance of the EMS, and to standardize optimized procedures wherever possible.
- Managing Change – one of the most challenging aspects of environmental management with respect to continuous compliance with applicable rules and regulations is the fact change can occur in virtually any area. Plant equipment or product formulations could

change, regulations may be amended, the applicability of rules to equipment could be changed due to new agency policies or other reasons, and so on. Thus, it is essential for an EMS to handle change effectively so that effort is at all times expended only to meet current, verifiable requirements in every unit, and that no regulatory requirement is lacking effort to document compliance. Managing change comprises three areas, all of which are extremely important to the success of an EMS. They are:

(a) recognizing that potentially relevant changes have taken place ("reconnaissance");
(b) determining how a change should be addressed in the EMS; and
(c) implementing the modifications to the EMS that are needed to ensure its continued effectiveness.

- Training – most compliance tasks will be the responsibility of staff members and, even in the case of "self-certified" tasks that rely on process historians to document compliance, human beings will need to supervise and verify results. Thus, the effectiveness of an EMS is ultimately a measure of the effectiveness of staff members when they carry out their assigned responsibilities.
- Automation – while some elements of a successful EMS in a large process plant literally require automation, such as the integration of process historians, many other aspects can be enhanced through the use of an automated approach. For example, the notification of compliance task "owners" when their tasks are overdue can be accomplished without computers and e-mails, however, (this leads to a large burden of "paperwork" with the potential for "human error.") Automation can improve efficiency, enhance reliability, and lead to a more predictable result, as compared to "paper-based" systems.

***"Raw Materials" for "Smart" Environmental Management Systems***

The inputs to a "Smart" EMS are summarized below:

- Equipment, Business or Operating Unit, and Site definitions (i.e., Name and Id. Number), including "mapping" of these entities to other pertinent asset identification systems (e.g., SAP, Preventative Maintenance, etc.)
- Applicability determinations – all applicable permit conditions and regulations must be defined for each piece of equipment or operating unit, ideally at the "citation" (detailed) level
- An EMS Outline that comprises all areas that will be managed, including regulatory compliance.
- Access to software that provides critical automated functions, such as notifying staff of overdue tasks or unexpected task completion results, maintaining equipment task lists, etc.
- Communications support systems that enable task "owners" to have access to supporting information, such as standard operating procedures, the full text of regulatory requirements, and other data.
- Adequate resources from plant managers and staff at all levels to obtain necessary approvals of EMS assignments, perform and record the completion of compliance-related tasks, and provide feedback to EMS administrators to support continuous improvement.
- Adequate support from the Information technology ("IT") departments to ensure the successful deployment and maintenance of the infrastructure and components required to operate the EMS, including web pages, e-mail notifications, etc.

- Access to information about sources of change that could affect applicable rules and tasks, such as from a focused subscription service or through review of professional publications.
- Access to process historian data and other relevant on-line data sources.

## A Blueprint for "Smart" Environmental Management Systems

The blueprint for a "Smart" EMS is different for companies with multiple operating facilities, as compared to companies that operate only a single plant. In the former case, experience has shown that efficiencies of scale can be gained by hosting the EMS database and system components at a central location, instead of hosting plant-specific databases and system components. In the latter case, a centralized system is not an option except with regard to hosting the databases and system components at a location remote from the plant location, such as at a "headquarters" office where IT resources may be more readily available.

"Smart" EMSs can benefit from hosting the EMS database and system components at a central location in a variety of ways. First, since there is only one set of software components and web pages for the enterprise, the effort required to perform routine updates and troubleshoot new issues is greatly reduced and simplified. Second, the web can provide highly available network resources to perform queries of plant-specific process historians, send e-mail notifications to users, and conduct other operations needed by the EMS. Since the web is essentially a "free" resource, its use to support efficient EMS operations is extremely cost-effective. Third, centralizing certain features and operations of the EMS enhances standardization of the EMS that can lead to faster enterprise-wide adoption of best management practices.

The use of generic information, i.e., information that can be propagated to many situations in the plant as opposed to "re-inventing the wheel," is another key aspect of the "Smart" EMS blueprint. EMS planners may make the mistake of over-emphasizing the differences in plants, units, equipment, and rules, and not taking advantage of reusing information where feasible in order to reduce effort and variability. For example, although the precise method to be used by Task Owner "A" to meet a regulatory requirement may differ from the method to be used by Task Owner "B" to meet the same requirement, the basic task to be performed is the same. The difference in methods does not support a strategy under which all elements of the compliance task assigned to Task Owners "A" and "B" should be different. An intelligently designed automated system will leverage common definitions of tasks, rules, and equipment, while accommodating true differences as needed to achieve the desired results.

When integrating on-line data sources such as process historians, with an EMS, it is important that this connection serve the purposes of: (a) documenting compliance data; (b); detecting non-compliance; and (c) notifying company staff when a problem arises. Notifications to staff should include not only potential deviations from applicable rules, but also notice of conditions such as "missing" data, server and network outages, and any time periods when the historian database is unavailable, such as during backups.

The EMS should store only the data needed to confirm compliance when it queries the historian database. For example, if the flow rate of water to an aqueous scrubber must be no less than 20 gallons per minute, and a flow meter is sampling flow and reporting it to the process historian every few minutes, it may suffice to record and evaluate for compliance the minimum flow rate in each 24-hour period in the EMS. Wherever possible, duplicative storage of data should be avoided.

The main feature of a "Smart" EMS blueprint is that EMS administrators should spend the majority of their time managing the system instead of managing compliance per se. To the extent that an administrator has responsibility to perform compliance tasks, then he or she is also a "do-er" in the ranks of personnel with task assignments. However, the principal role of the EMS administrator should be to ensure that:

(a) the EMS is up-to-date with current applicable rules and tasks, equipment status, and other system definitions; and

(b) task completion data compiled in the system has received proper attention from managers, and was reported to agencies where necessary. In summary, a "Smart" EMS should actively monitor a facility's compliance on a day-to-day basis, and informs decision-makers of information that they need to know.

## 201d Cloud Computing in Manufacturing Environment

**Son X. Huynh**, Global C&P Industry Solutions, IBM, 1475 Phoenixville Pike, West Chester, PA 19380 and Dennis A. Quan, High Performance On Demand Solutions (HiPODS), IBM, 555 Bailey Ave., Santa Teresa Lab, San Jose, CA 95141-1003.

"The impact of Cloud Computing in Manufacturing Environment"

Innovation is necessary to ride the inevitable tide of change in the today's manufacturing environment, especially in the information technology and automation landscape. Enterprise strive to reduce computing costs, to improve resource utilization and to achieve more efficient energy and environment use of their IT hardware and software investments. Cloud computing infrastructure accelerates and fosters these objectives by providing unprecedented flexible and dynamic IT resource pools, virtualization and high availability.

This paper introduces the value of implementing cloud computing and usage scenarios in manufacturing environment, especially in Chemical and Petroleum industries which typically have large numbers of heterogeneous applications, various hardware and huge data amount generated from sensors and devices in real-time and event-based exploration and production operations.

# AUTHOR INDEX

**American Institute of Chemical Engineers**
3 Park Avenue
New York, New York 10016-5991

**ISBN: 978-1-60560-218-9**